ON THE AIR

National Portrait Gallery

The Museum of Broadcasting

ON THE

Pioneers of American

by Amy Henderson

Published by the Smithsonian Institution Press
for the National Portrait Gallery
Washington City
1988

AIR
Broadcasting

Library of Congress Cataloging-in-Publication Data
Henderson, Amy.
On the air.
At head of title: National Portrait Gallery [and] Museum of Broadcasting.
"An exhibition at the National Portrait Gallery, October 7, 1988, to January 2, 1989"—T.p. verso.
Bibliography: p.
Includes index.
1. Broadcasters—United States—Biography—Exhibitions. 2. Entertainers—United States—Biography—Exhibitions. I. National Portrait Gallery (Smithsonian Institution). II. Museum of Broadcasting (New York, N.Y.) III. Title.
PN1990.7.H46 1988 384.54′092′2 88-15763
ISBN 0-87474-499-7

Cover illustration
Radio Talent (detail)
Watercolor on board by Miguel Covarrubias, 1938.
National Portrait Gallery, Smithsonian Institution
Illustrated in full on page 12
Key on foldout pages 201–202

Title page illustration
Jack Benny (detail)
Charcoal on paper by René Bouché, circa 1955. CBS Inc.

Jackie Gleason as the Poor Soul.

CBS Inc.

An exhibition at the National Portrait Gallery
October 7, 1988, to January 2, 1989

National Portrait Gallery, Washington, D.C.
Alan Fern, Director
Beverly J. Cox, Curator of Exhibitions
Suzanne C. Jenkins, Registrar
Nello Marconi, Chief, Design and Production
Frances Kellogg Stevenson, Publications Officer

The Museum of Broadcasting, New York City
Robert M. Batscha, President
Letty E. Aronson, Vice-President
Susan B. Fisher, Director, Capital Campaign
Douglas F. Gibbons, Director of Administration
Andy Halper, Curatorial Director
Ron Simon, Curator, Television
Karen Szurek, Director of Individual Giving and Membership
M. David Weidler, Treasurer

Contents

Foreword

Within the lifetimes of many of us, radio and television have come to serve as our most pervasive national media. Listeners and viewers receive programs broadcast from anywhere in the world, and Americans are used to the transmission of a rich diet of news, sports, entertainment, and cultural programs served up on an instantaneous basis.

From time to time, the National Portrait Gallery has explored various components of the American experience in its special exhibitions, but until now we have not investigated the media of communication, which have virtually bound together the United States into a single audience. In the political sphere the Gallery's exhibitions have included the Declaration of Independence, the American Revolution, and the Treaty of Paris, setting our nation on its course as an independent entity based upon newly defined concepts of freedom, human rights, and governance; and, more recently, "Adventurous Pursuits" celebrated the beginnings of diplomacy and trade with China. We have dealt with entertainment, in the broadest sense, in our exhibitions "Champions of American Sport," "Portraits of the American Stage," and "Hollywood Portrait Photographs," recalling people who have seized the imagination of the public by their talent, charisma, or skill.

Now we turn to the media of communication.

Even before Independence, the press played an important role in forming public opinion and rallying readers around a political or social viewpoint; early in the nineteenth century weekly and monthly periodicals addressed to the entire nation came into being, but American newspapers have tended to be local institutions, established to reach the citizens of a single community.

Broadcasting is so familiar today that few of us stop to reflect on what a technological miracle it is. It has reached such a peak of perfection that we are irritated when a program is interrupted during a storm or through a transmission failure at the station, and we take it for granted that when an event of national or world significance occurs we will learn about it without delay.

But not so long ago the world was dazzled simply by the possibility of transmitting the dots and dashes of Morse code without having to connect the sender and receiver by wires. That we would routinely receive sound in high fidelity and pictures in full color was inconceivable three-quarters of a century ago.

The development of radio and television is actually an international story. Guglielmo Marconi, inventor of wireless transmission, was Italian, and to this day there are fierce disputes about the contribution of innovators from other nations and about such issues as whether Great Britain or the United States can claim pride of place in the emergence of television broadcasting. What is certain is that broadcasting has had a uniquely unifying effect on the diverse population of the United States. I recall the several generations of my own family gathered around the radio in a Detroit living room, laughing together at the political and social satire of Will Rogers, anxious to hear the latest news from Hans Von Kaltenborn, or enjoying the Metropolitan Opera. My grandparents, who were recent emigrants from Europe, associated with the comedy and culture and information we received no less than did the younger, native-born Americans of the family, and their language and thought reflected in countless

Detroit broadcast of the 1922 World Series, featuring the New York Giants and the New York Yankees. WWJ was hooked to New York by telegraph for this event.

Detroit Historical Museum

ways what they had picked up from the programs we listened to. In my graduate-school days there was even a story of an immigrant professor from Hitler's Europe who innocently learned his idiomatic English from the radio serial "Abie's Irish Rose," with astonishing results that had to be corrected by his university colleagues.

As so often happens with something this familiar, few of us have stopped to reflect on how radio and television came to be such a powerful and ubiquitous force. In *On the Air* we meet the inventors and innovators, the people who devised the technical means for sending reproductions of human and musical sounds through the air, and the people who created the programs utilizing this new technology. Among these pioneers are those who found the way to link the various geographical areas of the country through the creation of networks, and those who created links between business and broadcasting so that programming could be made available without directly charging the listener; those who adapted existing forms of entertainment to the new medium, and those who invented new forms of programming that could never have existed before broadcasting did.

In this book, and in the exhibition related to it, historian Amy Henderson has selected some of the pioneers of the history of broadcasting around whom this story can be told. Her task has been immeasurably assisted by the knowledgeable staff of the Museum of Broadcasting, which has joined us in organizing this exhibition. All of us are deeply grateful to the Museum's director, Robert Batscha, and his colleagues for their enthusiasm and help. The Gallery appreciates the generosity of the many others who have assisted by lending portraits and objects, and by sharing insights. They are properly acknowledged elsewhere in the book, but I owe a special word of thanks to Frank Stanton, both for his service as a Commissioner of the National Portrait Gallery and for his support of this project from its outset.

For every pioneer included here, ten may have been passed over, but this is only because the group had necessarily to be limited in size. The history of radio and television is rich in its variety, and everyone will mourn the absence of some favorite inventor, performer, or innovator. Instead, I invite you to rejoice in the diverse array of personalities presented here, and to reflect on the story of the development of a modern phenomenon, in its peculiarly American form.

Alan Fern
Director
National Portrait Gallery

Preface

When the history of our century is written, it is probable that among the most significant events will be man's landing on the moon and the invention of both radio and television.

Few challenge the significance of man's landing on the moon, an event certainly no less important than Columbus's discovery of the New World five hundred years earlier. Similarly, the development first of radio and then of television has revolutionized the manner in which people communicate through time and space. The comparable milestone, of similar world-historic consequence, was Gutenberg's printing press, along with the development of movable type, again more than five hundred years ago. Just as we are at the very beginning of the space age and the excitement and challenge of new worlds of exploration and discovery, we are at the very beginning of a revolutionary means of communication among peoples and nations. We now have the capability to transmit the sounds and images of our civilization and culture from one generation to the next.

Think of the rich, creative history of the last five hundred years since the democratization of literature and communication made possible by the printing press; think of how books and periodicals have come to define our culture and our various civilizations, and the impact they have had on religion, politics, and political systems. For better and for worse, the world is very different as a result of Gutenberg's invention.

Now, imagine five hundred years into the future. What will historians conclude when they reflect on radio and television's influence on our literature, our religious practices, our politics, and our social and political systems? One thing is sure: radio and television's impact will be no less significant than Gutenberg's printing press. The men and women who are represented in this exhibition are the pioneers of a new era, the implications of which we can only begin to imagine.

While we cannot tell what the future might look like, we can speculate on how our perception of the past would have been different if these mediums of communication had been in use. Imagine television coverage of our country's first two hundred years. How much different and enhanced our understanding of historical events would be if we could witness George Washington crossing the Delaware or hear Abraham Lincoln delivering the Gettysburg address, watch the debates of the Constitutional Congress, or relive the trauma of Lincoln's assassination. Future generations, however, will be able to see the D-Day landing and Dr. Martin Luther King's "I Have a Dream" speech, watch the Watergate hearings, and relive the trauma of the Kennedy assassination. Imagine television coverage of Columbus's voyage to the New World. It sounds absurd, but television did go to the moon! Is there a person in human history we would not like to have on videotape? George Washington, Napoleon, Sir Isaac Newton, Jesus Christ, Moses, Confucius, Caesar—the list goes back to the beginning of time.

Most significant events and world leaders in human history from this century forward will be chronicled by these exciting new media.

Human culture, however, is more than historic events. It is how we live, how we express our perceptions of ourselves and the world we live in, the things we think about and laugh at, and how we spend our time. Television and radio will provide future generations with a new perspective and insight into what we were as a people: our everyday lives, how we were entertained, what we laughed at and what brought us to tears, what we valued, what we wore, how we lived, and how we changed. News reports will tell future generations what we thought was newsworthy and what we believed at the time was happening to us. Books and periodicals have already given us a richer understanding of the mores of the last five hundred years than of earlier cultures. Future generations will have that much more—the sound and sight of our times.

Finally, television and radio are generating the literature of our times. Today, some of our most creative people are working in these media to produce our children's folklore and the artistic milestones of our generation. As a record of performance, our century's legacy is enriched. For example, Gustav Mahler and Arturo Toscanini were contemporaries, and both were conductors of the Vienna Symphony Orchestra and the New York Philharmonic. We have only written records of Mahler's genius as a conductor, yet we have scores of radio recordings of Toscanini's conducting, as well as ten television recordings. How exciting it would be to have recordings of William Shakespeare's original casts performing his plays. Future generations will have, for example, productions of Arthur Miller's *Death of a Salesman,* one with Lee J. Cobb and another with Dustin Hoffman.

As one considers the portraits of broadcasting's creative community represented in this exhibition, one realizes what a rich legacy this first generation of broadcasters has created. The writers, comedians, actors, directors, and executives in this exhibition are our pioneers. In the decades and centuries to come, some will become our immortals, and their work the classics for future generations to emulate and build upon.

The significance of this exhibition is evident: it chronicles the beginning of a world-historic change, showing those who made it happen and how it was first used. But it is not simply an exhibition of broadcast pioneers. It is a portrait of us—as a people, as a culture, and as a civilization.

Robert M. Batscha
President
The Museum of Broadcasting

Radio Talent

Watercolor on board by Miguel Covarrubias, 1938. National Portrait Gallery, Smithsonian Institution

Key on foldout pages 201–202

Memoir

MARCONI . . . SARNOFF . . . PALEY . . . MURROW . . . BENNY . . . BALL . . . TOSCANINI . . . CRONKITE . . . GODFREY . . . HOPE . . . CROSBY . . .

One stands in awe. Not only of these unforgettables, but of the hundreds and thousands more who gave us their all "On the Air." Each was perfectly matched to his or her time. It is to the sum of these pioneers and their peers that the Smithsonian Institution and the Museum of Broadcasting have dedicated this exhibition, and to whom I dedicate these few thoughts.

Broadcasting is not that old, as media go. The commercial variety dates only from the 1920s, and some of its early pioneers—William S. Paley conspicuously among them—are still active in its vineyards. It is, they say, a young person's medium. It has helped keep some of us young who might have aged more rapidly at other labors.

Indeed, for us, no matter how hard the work, it was never labor. How could it have been, reinventing the universe each morning and reassembling it again at night? There was always a "First Nighter" quality about radio and television—to borrow from the title of a radio series in the thirties. To the extent that the broadcast media hope to continue to grow and prosper, they must never lose that quality.

It's easier to be a pioneer when you have to be. The men and women who gave us what we call the American system of broadcasting had one thing in common: much of the time, they didn't know what they were doing. Perhaps more importantly, they didn't know what not to do. But they did know what they *wanted* to do, so they just went about doing it, day after day and against the odds, and the better ones succeeded. That's harder to do today in broadcasting, although still a little easier in cable. The important thing is to remember that pioneering isn't reserved for the past, and is the only guarantor of a place in the future.

Broadcasting is show business, certainly, but not first and foremost an entertainment medium. It signed on with a serious pursuit—the Harding-Cox election returns, on KDKA Pittsburgh. It is for its real-world moments, then, that radio and television are longest remembered. It is the ability to juxtapose both the serious and the hilarious that distinguishes the electronic media from all others.

There is a great temptation to look back to the "Golden Years" and to treat the present with disdain. There was a "Playhouse 90," but there was no CNN. There was a "Gunsmoke," but there was no "L.A. Law." There was a "Mercury Theatre," but there was no "All Things Considered." (There still is a "CBS World News Roundup," happy to say.) But broadcasting, in all its manifestations, including cable and direct-broadcast satellite and high-definition television, has just begun to spread its wings.

Frank Stanton
President Emeritus
CBS Inc.

Acknowledgments

One of the great delights of this project has been working with so many people who have contributed their time and knowledge along the way. First of all have been the pioneers, many of whom made themselves available for interviews, while others provided the kind of primary chronicling that helped to shape this book and exhibition: Steve Allen, Fran Allison, Gene Autry, Lucille Ball and Gary Morton, Red Barber, Mr. and Mrs. Milton Berle, Mrs. Henry Cannon (Minnie Pearl), David Brinkley, Mr. and Mrs. Sid Caesar, Norman Corwin, Walter Cronkite, Mr. and Mrs. John Daly, Bob Elliott and Ray Goulding, Mark Goodson, Bob Hope and Mrs. Wyn Hope, Mr. and Mrs. Clayton Moore, Mr. and Mrs. Henry Morgan, Carlton E. Morse, William S. Paley, Reginald Rose, Robert Saudek, George Schaefer, Frank Stanton, John Cameron Swayze, and Sylvester Weaver.

The following families and associates of the pioneers were extraordinarily helpful: Miss Joan Benny, Mrs. Gioia Marconi Braga, Irving Fein, Dr. Thomas T. Goldsmith, Rolf Kaltenborn, Mrs. Lee Lawrence, Mrs. Dora Liberace, Franklin Mewshaw, Bernard Smith, Mr. and Mrs. Larry Spellman, Lowell Thomas, Jr., and Mrs. Rudy Vallee.

The research for "On the Air" was also carried out in numerous institutions. Here I was particularly lucky, for during the years of this project's gestation, many people who helped me became great friends as well as wonderfully instructive colleagues: at the Smithsonian's National Museum of American History, Carl Scheele and Ellen Roney Hughes in the Division of Community Life; John Fleckner, Robert Harding, Lorene Mayo, and Stacy Flaherty at the Archives Center; Elliot Sivowitch in the Division of Electricity; and Edith P. Mayo, Keith Melder, and William L. Bird in the Division of Political History. In the Smithsonian's Development Office, I would especially like to thank Bonnie Krench Miller, who worked tirelessly on fund-raising for the exhibition. At the Library of Congress, I should like to thank Robert Saudek, David Parker, Sam Brylawski, and Edwin Matthias in Motion Pictures, Broadcasting, and Recorded Sound; and Bernard F. Reilly, Jr., in Prints and Photographs. At the State Historical Society of Wisconsin, I am indebted to George Talbot, Maxine Fleckner Ducey, Barbara Kaiser, Myrna Williamson, and Christine Schelshorn. At the Broadcast Pioneers Library, Catharine Heinz has been a priceless resource, as has her assistant, Maria Luisa Llacuna. I would also like to thank the Pacific Pioneer Broadcasters, particularly Lenore Kingston Jensen. The Armed Forces Radio and Television Service Broadcast Center unleashed the full power of Mrs. Dorothy McAdam to help in my work on radio during World War II. Because of her and Ted L. Daniel at the Pentagon, I heard from people all around the world about radio's role in the war. And I am very grateful to the networks for allowing me to research their photo archives. At NBC I would like to thank George Rivera and Tony Page; at CBS, Louis Dorfsman, Martin Silverstein, and Kris Slovak.

I would also like to thank Himan Brown, Thomas A. DeLong, Margot Feiden, Howard B. Gotlieb of Boston University's Special Collections, Dr. Gene Gressley and Mrs. Paula McDougal of the American Heritage Center at the University of Wyoming, Robert Hupka, Carl Johnson of Williams College, Bruce Kelley of the Antique

The Quiz Masters

Lithograph by Al Hirschfeld, circa 1955. The Margo Feiden Galleries, New York City

Wireless Association, Margot Keltner of *Fortune* magazine, Kayla Landesman of UCLA's Manuscript Division, Stuart Lichtenstein of Sardi's, Martha Mahard at the Harvard Theatre Collection, Dwight M. Miller of the Herbert Hoover Library, Dr. Laura Montie and Roberta Zonghi at Boston Public Library, Richard Osgood, Walter Scharfe of the Brown Derby, Anthony Slide, David Smith of the Walt Disney Archives, Roy Stevens, Jeffrey Stewart, Devon Susholtz of the Humanities Research Center, Larry Viskochil of the Chicago Historical Society, Lew Wasserman of MCA, Don West of *Broadcasting* magazine, and Rosamond Westmoreland.

At the National Portrait Gallery, I first want to thank our Director, Alan Fern, who originally suggested the idea for this project and who then rarely interfered with it; I am deeply grateful to Marc Pachter, Assistant Director and Chief Historian, for allowing me time away from my regular duties so that I could pursue this project; and to Annetta McRae of the Historian's Office, for the incredible amount of work she dealt with so well. To my editors, Frances Stevenson and Dru Dowdy: you are the best—and thank you especially for your humor. I also am very grateful to designers Nello Marconi and Al Elkins, whose vision, flexibility, and talents made the exhibition come to life; to Wendy Reaves, Curator of Prints, and Bridget Barber; to William F. Stapp, Curator of Photographs, and Ann Shumard; to Cecilia Chin, NPG Librarian, Patricia Lynagh, and Martin Kalfatovic; to Suzanne Jenkins, Registrar, Jane Woltereck, Josette Cole, Eunice Glosson, Linda Best, Jack Birnkammer, and Andrew Wallace; to Chief Photographer Eugene Mantie and Rolland White; to the Smithsonian's Women's Committee for a yearlong fellowship which allowed me the brilliant help of Susan Smulyan in this project's early stages; to Claire Kelly of the Exhibits Office; and to interns Carol Dyer, Maya Arai, George Glastris, and Clare Franklin.

I am most grateful for the enthusiastic partnership provided by Dr. Robert M. Batscha and his staff at the Museum of Broadcasting, including Ron Simon, Letty E. Aronson, and M. David Weidler. The cosponsorship of the Museum of Broadcasting—and the creative energy of its staff—have been vital to the success of "On the Air," and all of us at the Portrait Gallery are extremely happy for this association.

I would also like to give special thanks to the following people: Erik Barnouw, Charles Jahant, John Meehan, Charles M. Sachs, and Walter Zvonchenko. And finally, there are two people to whom I owe an especial debt, for without them this project simply would not have been realized: Frank Stanton, National Portrait Gallery Commissioner and President Emeritus of CBS, and Beverly Jones Cox, Curator of Exhibitions at the Gallery.

NBC radio show

Broadcast Pioneers Library

Lenders to the Exhibition

Edgar Bergen Family

Mr. and Mrs. Milton Berle

Leonard Bernstein

Mrs. Gioia Marconi Braga

Broadcast Pioneers Library, Washington, D.C.

Sid Caesar

CBS Inc., New York City

Chicago Historical Society, Illinois

Norman Corwin

The Country Music Hall of Fame and Museum, Nashville, Tennessee

Virginia Warren Daly

Thomas A. DeLong Mighty Music Box Collection

The Margo Feiden Galleries, New York City

Fred Flowerday

Mark Goodson

Hamilton College, Clinton, New York

Harvard Theatre Collection, Cambridge, Massachusetts

Bob Hope

Lee Lawrence

Alice Becker Levin

Liberace Museum, Las Vegas, Nevada

Library of Congress, Washington, D.C.

Henry Morgan

Carlton E. Morse and Trustees, Morse Family Trust

Museum of Science and Industry, Chicago, Illinois

National Museum of American History, Smithsonian Institution

William S. Paley

Bob Paquette Microphone Museum, Milwaukee, Wisconsin

Private collections

Harry Ransom Humanities Research Center, Austin, Texas

Reginald Rose

George Schaefer

Anthony Slide

Bernie Smith

Society of Illustrators Museum of American Illustration, New York City

Frank Stanton

State Historical Society of Wisconsin, Madison

Tufts University, Medford, Massachusetts

University of California at Los Angeles

Mrs. Rudy Vallee

Sylvester Weaver

Wesleyan University, Middletown, Connecticut

Williams College, Williamstown, Massachusetts

Introduction: The Rise of the Middlebrow

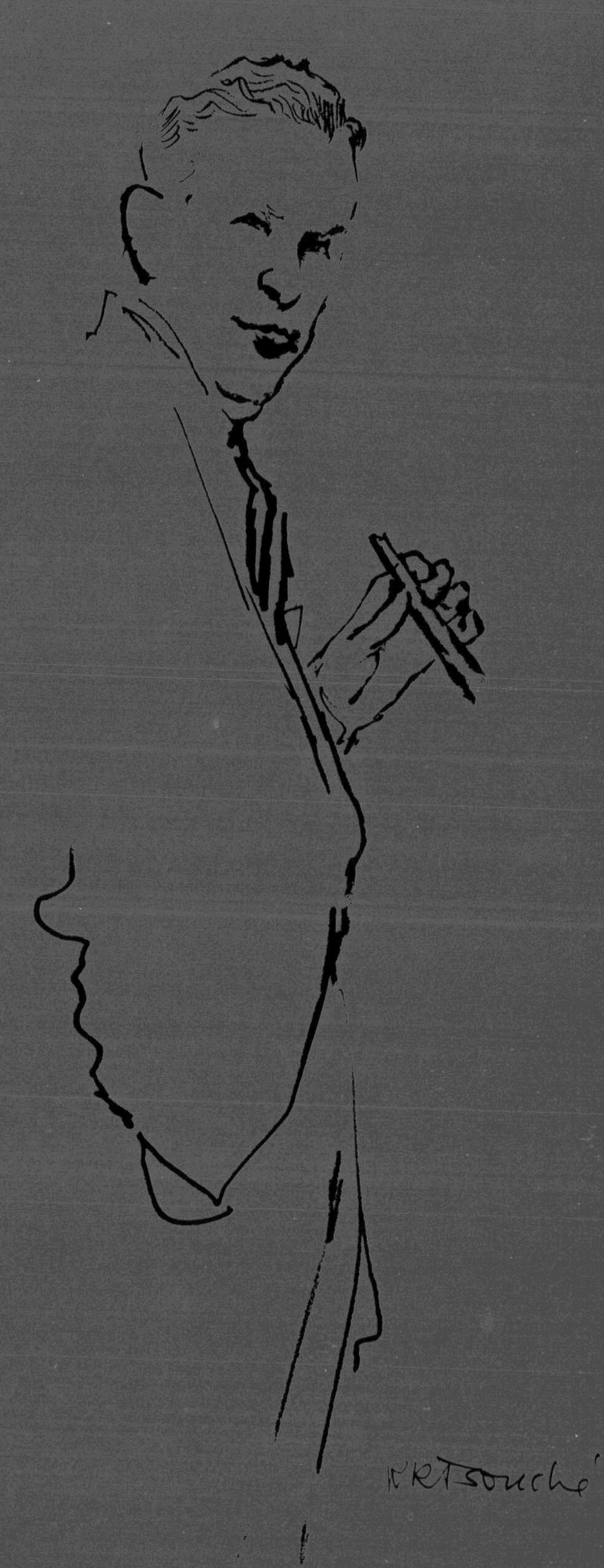

In the beginning of broadcasting, there were no words—only dots and dashes. Guglielmo Marconi's 1901 transatlantic transmission of the Morse letter "S" heralded a cultural and technological new age that paralleled America's own emergence as an urban, industrial society.

The idea of broadcasting as we know it—of airwaves resonant with news, music, drama, and entertainment—had to await an increased sophistication in technology. Until after the First World War, the "ether" was mainly the province of ship-to-shore naval communications. But by the late 1920s, following an explosive evolutionary burst, the dots and dashes of "wireless telegraphy" had metamorphosed into living, breathing radio. With stunning speed and few backward glances, the American broadcasting industry in the 1920s emerged in its commercial network form, thereby setting the stage for the most epochal change in everyday life since the Gutenbergs' movable type had begun the information revolution in the fifteenth century.

Early broadcasting was as much a product of an intellectual climate as it was of technology. In 1915 Van Wyck Brooks published *America's Coming of Age*, a critical look at the gap between traditional American culture and the social realities of the twentieth century. Lamenting the nation's lack of cultural integration, Brooks decried a society that still perceived itself as divided between "highbrow" and "lowbrow": "What side of American life is not touched by this antithesis?" The problem was that little attention had been devoted to creating a culture that accurately reflected the exigencies of a modern industrial society. As a result "there is no community, no genial middle ground." The hard fact of modern life, Brooks argued, was that American society no longer resembled an Edenic paradise; it was more like the Sargasso Sea: "All manner of living things are drifting in it . . . gathered into knots and clotted masses." Brooks's objective was to have America reexamine its view of culture and reality, to draw itself into the twentieth century by creating a new national culture, "which radiates outward and articulates the entire living fabric of a race," including the "middle tradition."

The answer to Brooks's criticism was a machine-age culture—an aesthetic of industrial optimism—which emerged in the 1920s and 1930s. The beauty of this solution was that it fit into an established historical debate over the role of technology: would technology produce an ideal social order, or would it lead to a mechanized, soulless society? Rooted in the Enlightenment and Jeffersonian thought, the dispute began when early industrialization came into conflict with eighteenth-century notions about the potential of the unspoiled New World to renew and regenerate mankind. In *Notes on the State of Virginia* (1787) Thomas Jefferson advocated America as a nation of middle-class yeoman farmers. "It is the manners and spirit of a people," he wrote, "which preserve a republic in vigour." Industrial society spawned "mobs of great cities" that were "sores" on the body politic: "Let our work-shops remain in Europe," he argued, and let America be peopled by yeomen. "Those who labour in the earth are the chosen people of God."

Opposed to Jefferson's agrarian ideal were those who argued that only through "the mechanical arts" could mankind improve its situation. Industrialization, to such Jeffersonian contemporaries as Charles Willson Peale, was the means to the ideal middle landscape between barbarism and over-civilization, freeing man from drudgery and bringing about the betterment of all society. This argument prevailed in the course of the nineteenth century, with industrial progress eventually becoming a barometer by which the national worth itself was gauged.

The implications of harboring what historian Leo Marx classically called "the machine in the garden" remained long after the rural landscape receded. The foreboding intellectual atmosphere in the aftermath of the First World War exacerbated the debate over technology in the 1920s and 1930s: would man be the servant or master of the machine? The anti-machine contingent argued that technology represented an ominous *deus ex machina*, which threatened to dehumanize society. The denizens of Fritz Lang's 1927 silent film, *Metropolis*, were depicted as robotic masses, while Busby Berkeley—in the 1928 "Earl Carroll's Vanities" and later in Hollywood films—used look-alike dancers as if they were interchangeable parts in vast machine-dance production numbers. The protagonist of Elmer Rice's 1929 play, *The Adding Machine*, was named "Mr. Zero"; Rice characterized Mr. Zero as a virtual machine cog. In 1932 Aldous Huxley fictionalized a dark *Brave New World* in which the planetary motto was "community, identity, stability"; elsewhere Huxley railed against "the talkies" as "the latest and most frightful creation-saving device for the production of standardized amusement."

The "machine romantics," on the other hand, celebrated technology for its potential to bring a new harmony to industrial life. In *Horizons* (1932) Norman Bel Geddes viewed the machine culture as a merging of aesthetics and rationality. Against charges that machine-age culture would lead to a standardized and hopelessly uniform society—a proto-Fascist state—Sheldon and Martha Cheney argued in *Art and the Machine* (1936) that machines were "a new tool epochally brought to man's service." Everyday machines promised man freedom from the minutiae of daily routine.

The focus of machine-age culture, especially in Depression America, was on social redemption and order. As historian Warren Susman wrote in an essay called "Culture and Communications," "By the 1930's, the . . . expert human designer had

Me and the Set

Lithograph by Al Hirschfeld, 1955. The Margo Feiden Galleries, New York City

in a sense replaced the eighteenth century vision of God as a god of design. In a world increasingly out of order . . . man as designer was called upon to find some new order in the world."

One of the most important results of the machine-age aesthetic was the mass production of culture. Movies became part of neighborhood life; even during the Depression, eighty million people a week went to "the pictures." The recording industry flourished as well, with RCA Victor reporting that record sales increased by 600 percent between 1933 and 1938. But it was radio that became the archetypal medium for the mass production of culture. Unlike movies, radio was a household presence: in 1934 an average radio cost $34.65, and 60 percent of all American homes had at least one set. And, unlike records, radio was live: entertainment and information were there at the touch of the dial. Whether or not radio ever proved to have what NBC president David Sarnoff envisioned as a "gift for democratizing," it is clear that the medium achieved a far more vital—and less morally freighted—function simply by providing a culture that was accessible.

This was to be radio's greatest contribution to the national ethos; in fact radio's gift *was* a national ethos. The formative years of broadcasting coincided with America's own tilt toward modernity. An urban, industrial society replaced the largely rural landscape, and millions of new immigrants supplanted the nineteenth century's dominant genteel tradition with a kaleidoscopic ethnicity. From the shards of diversity and change, radio provided what historian Carl Scheele has called the "cultural glue" that created a new commonality—a truly national culture—in the late 1920s and 1930s.

Television brought an entirely new dimension to broadcasting, one far beyond the simple addition of "sight to sound" that David Sarnoff announced at the 1939 New York World's Fair. In the first era of commercial television, between 1948 and 1960, the small screen posed the same questions about mass-produced culture that radio had raised a generation earlier. The role of technology again became a central point of discussion, focusing especially on broadcasting's responsibilities as a mass medium.

A new factor introduced by television, however, was its literalness. Unlike radio, television left little to the imagination, and everything seen through its eye—drama, variety, news, politics—took on a documentary credibility. As Daniel Boorstin has argued in *The Image*, television created its own "pseudo-events" that blurred distinctions between fact and fiction. Was the televised image real? Was it ersatz? Did it matter? Did this blurring lead to the creation of homogenized stereotypes, repeating so-called facts until they had taken on the appearance of truth? Had technology brought about a new level of the "hieroglyphic civilization" that Vachel Lindsay described in his pioneering 1915 study, *The Art of the Moving Picture*?

And what did an increasingly visual and literal culture portend? Did the form and function of television shape American culture, or did American culture—as Warren Susman has suggested—shape the form and function of TV? Whereas radio had been a unifying force, creating a national culture out of diversity, did commercial television fragment its mass audience, as indicated by the growth of FM radio and educational TV? Were the media's rays finally able to achieve an Aristotelian balance, a cultural middle ground bringing the greatest good to the greatest number? Did television bridge the gap that Van Wyck Brooks had disparaged between highbrow and lowbrow culture? Was television's mass cult in fact the ascension of the middlebrow?

This book is a cultural history rather than a history of broadcasting, and it focuses on the people who created the industry rather than on the process itself. The pioneers of broadcasting highlighted here are a representative instead of a definitive selection, chosen not only for their particular contributions, but because they also illustrate well the kinds of cultural issues that have most fascinated me.

Beginning with turn-of-the-century inventors, *On the Air* is organized by such genres as drama, news, and variety; a topical approach was used because broadcasting's major personalities often had enduring careers that spanned both radio and television, and because broadcasting's programming formats changed very little from early radio until the end of TV's first decade. Once broadcasting found its basic rhythm in the 1920s, it barely missed a beat during the next few decades, even with the advent of television.

On the Air closes with the 1960 Kennedy-Nixon debates, a transitional moment that marked the end of television's first era—predominantly live and in black-and-white—and signaled the medium's emergence as a central presence in American life. The implications of this presence are as ubiquitous as broadcasting itself—although the medium's message, frankly, provokes many more questions than it provides answers.

Inventors and Entrepreneurs

The pioneers of broadcasting reshaped American life and culture literally out of thin air. In their creation of a commercial network enterprise, these inventors and entrepreneurs forged a national system of mass communications. As one writer observed in *Collier's* magazine in 1922, radio broadcasting was "something entirely new," capable of bringing "entire nations into one vast audience, to make hundreds of millions hear the same thing at the same time." More than any other agency, radio had the potential for "spreading mutual understanding to all sections of the country . . . unifying our thoughts, ideals, and purposes . . . making us a strong and well-knit people."

Radio was hailed with utopian fervor in an age when technological progress was embraced as the panacea for mass industrial society. Those who contributed to that progress became national heroes, as celebrated for taming urban diversity as earlier generations had been for subduing the wilderness.

In a 1928 article in *The Forum* on "Machinery, the New Messiah," hero-inventor Henry Ford wrote that "Machinery is accomplishing in the world what man has failed to do by preaching, propaganda, or the written word." What radio's pioneers did was to give the machine culture a voice.

David Sarnoff and Guglielmo Marconi at the RCA Communications transmitting center in Rocky Point, Long Island, in 1933. Sarnoff started his career as an office boy for Marconi.

NBC

Guglielmo Marconi.

Oil on canvas by Gustave Muranyi, 1914. Mrs. Gioia Marconi Braga

Guglielmo Marconi 1874–1937

The real genius of the "Wizard of Wireless" lay in his ability to harness the brave new technology of wireless communication to the public imagination. By popularizing wireless and making it both practicable and commercially feasible, Guglielmo Marconi rooted himself in the tradition of Samuel F. B. Morse, Alexander Graham Bell, and Thomas Edison: larger-than-life figures whom Americans elevated into a pantheon of hero-inventors.

Born the second son of a wealthy Italian landowner, Marconi spent his youth on his father's estate in Bologna. By the time he was twenty, he was experimenting with the electromagnetic waves discovered by Heinrich Hertz. No one knew how to control these waves until Marconi began his experiments. Using his family as an audience, he demonstrated how he could ring a bell across a room with radio waves—a feat his barely amused father considered a parlor trick. Then Guglielmo discovered that ever-greater distances could be achieved by using larger and larger antennas, with groundings at both the transmitter and the receiver. Using the dots and dashes of Morse code, he was eventually able to transmit over the mile-and-a-half length of his family's estate; at this point, even his skeptical father approached enthusiasm.

Marconi's magnetic detector, circa 1902–1912, which he refined for use in both ship and shore installations. The "maggie" had been developed to improve detection of early wireless telegraph signals.

National Museum of American History, Smithsonian Institution

The family contacted the Italian government, but the Minister of Post and Telegraph rejected Marconi's system as a duplicate of the already-existing telegraph system. Guglielmo and his mother then went off to England, where Marconi amazed the public with demonstrations of ship-to-shore transmissions through rain and pea-soup fog, and over long distances. As a result, the British navy began to equip its warships and lighthouses with his wireless. By 1899, enough international eyebrows had been raised that James Gordon Bennett, editor of the *New York Herald* and commodore of the New York Yacht Club, offered Marconi $5,000 to come to America to transmit the America's Cup Race. By getting Marconi's minute-by-minute account of the race, the *Herald* figured to have the results in print before the yachts even returned to shore. The experiment worked, and a month later, "American Marconi" was launched with an authorized capital of $10 million.

Marconi was custom made for an age that apotheosized scientific progress. With the exception of a few old worriers like Henry Adams, who lamented—in his autobiography, *The Education of Henry Adams*—the ascendancy of the "Dynamo" over the "Virgin," most Americans venerated those who worked to control Nature for the good of mankind. According to scientific periodicals of the day, hadn't the "Century of Progress" just culminated in "an epidemic of industrial progress"? Wasn't distance being relentlessly annihilated and the planet reduced to a "tiny satellite"?

Marconi's next experiments exploited scientific progress across time and space. Setting up his instruments in Newfoundland on a cold December day in 1901, he successfully received the letter "S" relayed in Morse code from a transmitter 1,700 miles away in Cornwall, England: it was the first transatlantic communication by wireless telegraphy. The twentieth century had arrived with three short buzzes—the Morse code "S"—and Marconi was instantly famous.

Lionized now by the press, Marconi capitalized on the popularity of wireless. His special wizardry lay in seizing embryonic scientific discoveries and molding them into a commercially successful system of communication. The chief patents in his name were on improved types of antennas, coherers, magnetic detectors, and methods of selective tuning. In each case, though Marconi had not done the fundamental experimenting, he had provided the leap of imagination to find practical ways to apply these inventions to wireless. He was also one of the first to realize that "broadcasting" was an entirely new form of communication.

Yet he never made the ultimate leap: in these years he never envisioned a radiobroadcast industry based on voice transmission. For him the dots and dashes of Morse code were enough to fill the air.

The "wizard of wireless."

National Museum of American History, Smithsonian Institution

Lee De Forest 1873–1961

Determined to "send the human voice through the air instead of messages by dots and dashes," Lee De Forest invented the device that made voice transmission possible. Of all those associated with radio's development, he was virtually alone in envisioning broadcasting as the main use of wireless telegraphy and in foreseeing "the day when by the means of radio, opera may be brought into every home. . . . The news, and even advertising, will be sent out to the public on the wireless telephone." In experiments undertaken between 1905 and 1907, the "father of radio" ultimately fashioned a bit of wire and a lamp from a Christmas tree into an "audion," a three-element vacuum tube that has been called one of the fundamental inventions of the twentieth century. He had, as he later wrote, discovered "an Invisible Empire of the Air."

Lee De Forest, the "father of radio."

Bronze by Joseph Dominico Portanova, 1953. National Museum of American History, Smithsonian Institution

Born into a parson's family in frontier Iowa, De Forest grew up in a profoundly religious household and never lost a certain evangelical fervor. He earned a Ph.D. from Yale and worked for Western Electric before starting the series of experiments that led to the formation of the De Forest Wireless Telegraph Company in 1902. Here De Forest found himself in direct competition with the man he considered his nemesis, Guglielmo Marconi. His animosity was such that, in 1901, after Marconi had trumpeted his successful transatlantic reception of the Morse code "S," De Forest snidely recorded in his diary that "Signor Marconi has played a shrewd *coup d'etat*, whether or not the 3 dots [the Morse code "S"] he says he heard came from England." What especially piqued De Forest was that Marconi had "taken the thunder from competitors, who may in a few years actually send messages across the ocean. His stock is soaring, and will make the achievements of others, however meritorious, look cheap enough in the *popular* eye."

But De Forest didn't shy away from self-promotion, either. At the 1904 Louisiana Purchase Exposition in St. Louis, his showmanship glittered from the "De Forest Tower," a wireless station perched one hundred feet above the ground in a glass house. People flocked to the tower, drawn by what De Forest cheerfully described as the "staccato crackle of our spark, [which] when purposely unmuffled, brought them swarming from all over that end of the Exposition grounds." It was the sort of performance P. T. Barnum would have relished.

Back in his New York aerie on the top floor of the Parker Building, De Forest continued to attract popular attention. Using gramophone records, he broadcast "sweet melody . . . over the city and sea so that in time even the mariner far out across the silent waves may hear the music of his homeland." In 1908 he broadcast from the Eiffel Tower to a wireless operator 550 miles away in Marseilles; the next step would be a transatlantic broadcast.

Yet De Forest's fervor failed to keep him from severe business and financial problems, and a wary public viewed some of his stunts as mere chicanery. Part of the problem was the public perception of radio itself—this invisible, all-invasive ether. In 1906 the *New York Times* ran an editorial headlined "A Triumph But Still a Terror" and quoted one authority who shivered at the potential "babel of voices" and the specter of long-distance eavesdropping: "One could be called up at the opera, in church, in our beds. Where could one be free from interruption?" Where indeed?

To try to ameliorate this distrust, De Forest decided to show off one of radio's greatest assets: its ability to broadcast an event as it happened, to make that moment instantly accessible to whomever tuned in. On January 13, 1910, he set up two microphones along the footlights on the stage of the Metropolitan Opera. The broadcast that night of Enrico Caruso and Emmy Destinn in *Cavalleria Rusticana* and *Pagliacci* was not an unqualified success. The *New York Times* complained that, while the "worbling" of Caruso and Mme. Destinn was "not clearly audible," a member of the audience was distinctly heard to say, "I took a beer just now, and now I take my seat."

Although the experiment may not have been the complete triumph he sought, it marked the beginning of modern radio, and De Forest enthused that radio broadcasting in a few years would be "a medium for entertainment that might rival the stage, opera and concert hall." Caruso's "worblings" had established the potential of live broadcasting.

Over the next two decades, radio would develop much as De Forest had prophesied. But in one vital way—the commercialization of the airwaves—radio's path would diverge sharply from that outlined by its inventor. In years to come, De Forest would vociferously denounce the "etheric vandalism of the vulgar hucksters" who had traduced his self-styled "child," and "who, lacking awareness of their grand opportunities and moral responsibilities to make of radio an uplifting influence, continue to enslave and sell for quick cash the grandest medium which has yet been given to man to help upward his struggling spirit."

The particular form that American broadcasting took in the 1920s—anchored in a commercial network system—ordained that this would be one ethereal campaign that even De Forest's missionary zeal could not overcome.

De Forest broadcasting over station W2XCD, Passaic, New Jersey.

Photograph by Underwood and Underwood. National Museum of American History, Smithsonian Institution

Herbert Hoover 1874–1964

The regulation of the burgeoning radio industry in the 1920s was the responsibility of Herbert Hoover. As secretary of commerce in the Harding and Coolidge administrations, he oversaw the explosion of radio from the realm of amateurs with crystal sets to a national commercial network system.

Wireless had been put under the purview of the Commerce Department in 1912, before broadcasting was much more than the dots and dashes of ship-to-shore communications. Hoover recalled that his own son, after the First World War, had "gone daft on wireless" and the house had been cluttered with a variety of apparatuses. About six months after Hoover had become secretary of commerce, "suddenly a great public interest awoke in radio," and soon there were about 320 broadcasting stations, mostly of low power and short range.

The law regulating radio, Hoover said, was "a very weak rudder to steer the development of so powerful a phenomenon . . . especially as it so rapidly developed over the next few years." In 1922, to establish some structure giving order to the growing chaos, Hoover called a conference for all radio people—broadcasters, manufacturers, representatives of the army and navy, and the amateurs. One major problem was the plethora of radio receiving sets: their "comparative cheapness . . . bids fair to make them almost universal in the American home." Stating that, since it was "perfectly hopeless" to expect the use of the radiotelephone for communication between individuals, the focus of regulation needed to be on broadcasting—"the necessity to establish public right over the radio bands." Certain bands would be set aside for the army and navy, certain for public service, and one for the amateurs. Hoover also gave his assessment of the commercialism creeping over the ether: "It is inconceivable that we should allow so great a possibility for service and for news, for entertainment and education and for vital commercial purposes to be drowned in advertising."

As the radio industry continued its rapid expansion, Hoover called additional radio conferences, in 1923, 1924, and 1925, and further sketched out the skeletal framework of broadcasting regulations. His laissez-faire attitude about the role of the federal government had a formative influence on the extent of this regulation. An advocate of self-regulation, Hoover strongly objected to the sort of government monopoly on broadcasting that the British had established, arguing that "free speech and general communication would be safer in private hands." He also strongly opposed any private monopoly, saying that it "would be in principle the same as though the entire press of the country was so controlled." He still objected to advertising, though admitting that it was the only feasible alternative to government monopoly. It was almost an aesthetic problem for Hoover: imagine, he asked, if a presidential speech were ever used "as the meat in a sandwich of two patent medicine advertisements."

Hoover's conferences started a domino process that in 1927 led Congress to enact regulatory legislation. The Radio Act fostered the development of a national commercial broadcasting system based on the public ownership and regulation of wave channels. A Federal Radio Commission was set up to administer the law.

The 1920s were determinative for American radio, as the broadcast industry grew up to fill out the structure that Hoover had prescribed—"free of monopoly, free in program, and free in speech."

As secretary of commerce during the Harding and Coolidge administrations (1921–1929), Herbert Hoover established the laissez-faire federal policy that allowed a private commercial broadcast industry—rather than a government monopoly—to take root and flourish.

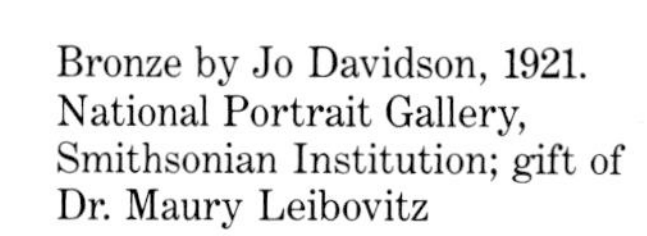

Bronze by Jo Davidson, 1921.
National Portrait Gallery,
Smithsonian Institution; gift of
Dr. Maury Leibovitz

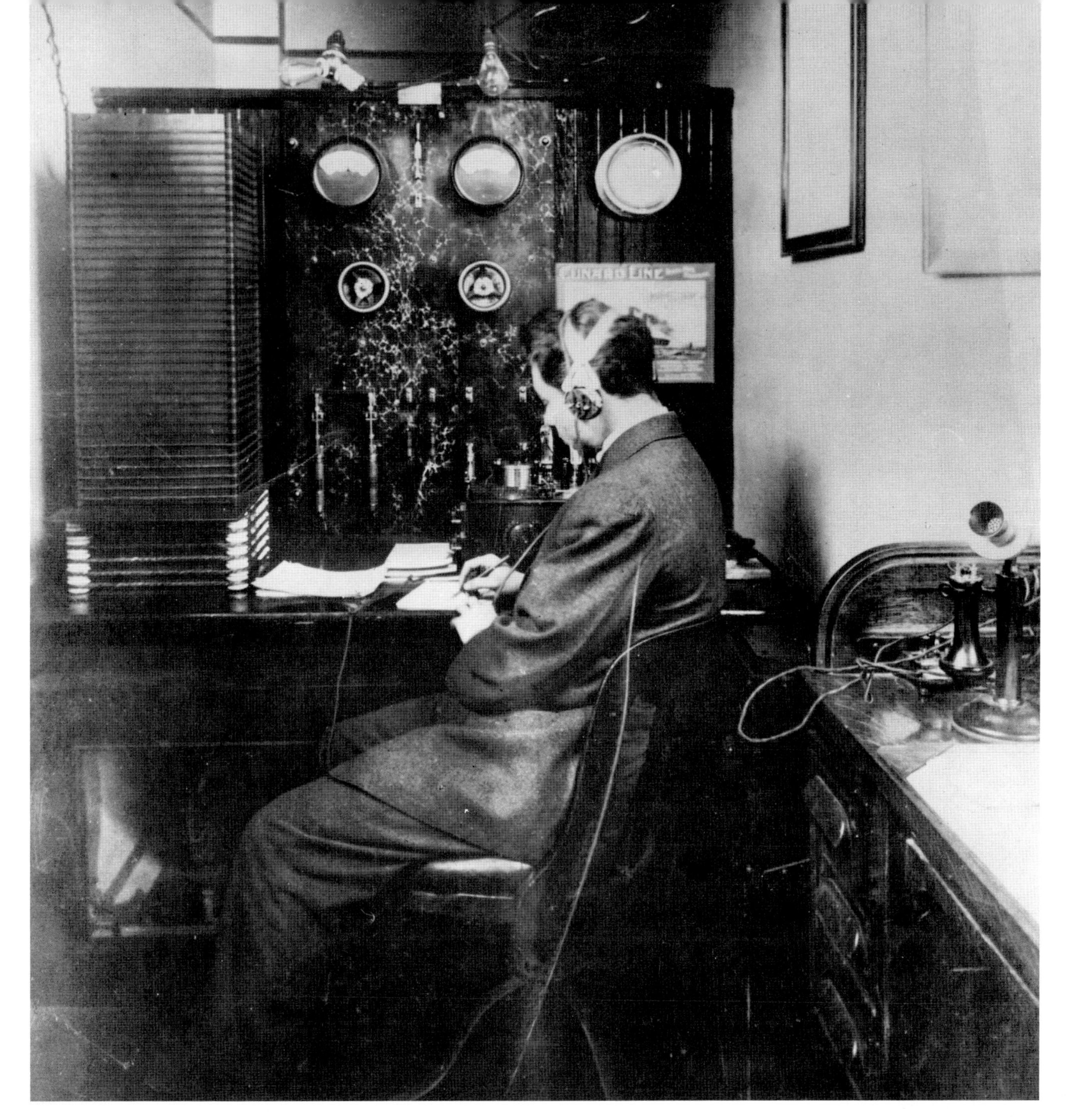

David Sarnoff on duty at the radio station atop the Wanamaker store in New York, 1912, reporting on the Titanic *disaster.*

RCA

David Sarnoff 1891–1971

David Sarnoff's life paralleled the emergence of an urban, industrial America, and, as both the "prophet of radio" and the "father of television," he helped forge the direction of modern life. Born in a poor village in southern Russia, Sarnoff immigrated to New York via steerage, and, as he once wrote, "Two days later I was peddling papers in the streets . . . to help support my family." He soon saved enough money to buy a telegraph key, learned the Morse code, and began working for the Marconi Wireless Telegraph Company of America.

As the wireless operator in the Marconi-owned station at Wanamaker's New York department store, Sarnoff was on duty on April 14, 1912, and was among the first to pick up a dim signal from the S.S. *Olympic,* 1,400 miles away: "*S.S. Titanic* ran into iceberg. Sinking fast." For the next seventy-two hours, Sarnoff stayed at his wireless board and relayed reports of the unfolding disaster to the nation through his exclusive reports to the *New York American.* The tragedy, he later recalled, "brought radio to the front, and incidentally me."

Whether or not Sarnoff, as legend has it, was the sole wireless operator involved in this incident, or, as recent research suggests, he was but one of many, the point is that the publicity resulting from the *Titanic* disaster vastly enhanced the position of American Marconi, and that of David Sarnoff as well. In quick succession, he rose through the company to become chief inspector for all ships equipped and serviced by American Marconi, and it was from this position in 1916 that he wrote one of the most visionary memoranda in broadcasting history.

Radio in the mid-teens had not yet found its voice: it was basically a system of ship-to-shore communication by dots and dashes. Yet Sarnoff foresaw a plan that "would make radio a 'household utility' in the same sense as the piano or phonograph." His "Radio Music Box" could be "placed in the parlor or living room," where listeners "could enjoy concerts, lectures, music, recitals, etc., which may be going on in the nearest city within their radius." But of even more potential, he continued, "Events of national importance can be simultaneously announced and received. Baseball scores can be transmitted in the air."

The management of American Marconi pronounced his idea "harebrained" and simply ignored the memo. But in 1919, when the company was absorbed into the Radio Corporation of America, Sarnoff marked his time and tried again. After the war the climate had changed, and within a year of RCA's founding, domestic wireless developments had reached a frenzy; crystal sets and ham radios made voice transmission the newest fad. Sarnoff now proposed his "Radio Music Box" plan to Owen Young, head of RCA, and in a follow-up memo estimated—in figures so prescient that they cemented his reputation as radio's prophet—that during the first year of production, more than 100,000 Radio Music Boxes would be sold at seventy-five dollars each; in the second year 300,000, and in the third 600,000.

Sarnoff also felt that RCA had to make its broadcast presence known in a dramatic way. Borrowing a portable transmitter, he arranged the July 1921 broadcast of the heavyweight-championship prizefight between Jack Dempsey and Georges Carpentier. Announced at ringside by Major J. Andrew White—with Sarnoff at his elbow—the four-round broadcast was a triumph, listened to by an estimated 300,000 people.

The years between 1922 and 1926 proved the most vital in American broadcasting, for it was in these years that the commercial network matrix was configured. There were more than six hundred radio transmitters operating nationwide in 1923, yet broadcasting itself remained essentially local. With the burgeoning of transmissions—including those from tens of thousands of amateurs who put their own sets together and merrily transmitted from garages and attics—the concern was that, with signals and static colliding across the unregulated spectrum, radio would likely drown in its own cacophony. Sarnoff recognized that some overseeing structure had to be imposed. He was the first industry leader to devise a plan for a national broadcasting company, which would be "established as a public service . . . [and] will ultimately be regarded as a public institution of great value, in the same sense that a library, for example, is regarded today."

The biggest obstacle to such a plan was removed in 1925, when Sarnoff negotiated American Telephone and Telegraph's withdrawal as a major radio competitor. AT&T would henceforth cease its broadcasting career, and RCA would lease the company's interconnecting telephone lines for $1 million annually. On September 9, 1926, the National Broadcasting Company was incorporated, promising in full-page ads to provide "the best program[s] available for broadcasting in the United States." Two months later the NBC Red and Blue networks were launched with a simultaneous four-hour broadcast. At New York's Waldorf Astoria Hotel, Walter Damrosch conducted the NBC Symphony, while Mary Garden sang "Annie Laurie" from Chicago, and Will Rogers did his Calvin Coolidge impersonation from Independence, Missouri. As the *Washington Post* reported, "Radio . . . has put aside its swaddling clothes and

David Sarnoff at the 1939 New York World's Fair announcing that the "sight" of television had been added to the "sound" of radio.

Barnouw Collection, National Portrait Gallery, Smithsonian Institution

An early visionary of broadcasting's potential, David Sarnoff became the guiding force of NBC in that network's formative years.

Charcoal on paper by S. J. Woolf, 1941. National Portrait Gallery, Smithsonian Institution

has become a potential giant."

The second phase of Sarnoff's career dealt with television. Initially interested because of Dr. Vladimir Zworykin's invention of the iconoscope in 1923, Sarnoff set up a special NBC station (B2XBS) in 1928 to experiment with television. The first public demonstration of television came at the opening of the 1939 New York World's Fair, when on April 30 Sarnoff himself was televised, proclaiming, "Now at last we add sight to sound." World War II interrupted television's smooth emergence until the late 1940s, when once again Sarnoff was in the forefront, leading the way for a compatible system of color television.

In his 1939 World's Fair speech, Sarnoff had emphasized broadcasting's "creative force which we must learn to utilize for the benefit of all mankind." The uplifting potential of broadcasting was one very near the heart of the immigrant from southern Russia. It was Sarnoff's belief in broadcasting as a cultural agent that led him to initiate broadcasts of the "Music Appreciation Hour" with Walter Damrosch, the Metropolitan Opera, and, in perhaps his greatest coup, the NBC Symphony under Arturo Toscanini. He firmly believed that radio—and later television—had a "gift for democratizing" because "the dweller on the lonely prairies, and the farmer . . . could be given the opportunity to listen to and enjoy good musical composition . . . in the same way in which the rich man could do so by being in his exclusive box."

By fighting for this kind of broadcasting, Sarnoff made a major contribution to the cultural remelding of modern America. He had begun his career in an America fragmented by growing urbanization, immigration, and technological revolution. His greatness lay in seeing radio as a virtual melting box of disparate cultures, creating, in the 1920s and 1930s, a new common fund of experience and information that was democratic in its touch-of-the-dial accessibility. Because of his vision, broadcasting not only countered the disintegrative forces of a changing America, but in the process created a new national consciousness.

William S. Paley born 1901

"In the 1920s and 30s with radio, and again in the 1940s and 50s with television, CBS was doing more than building a business," William S. Paley said in 1983. "We were inventing a future." In 1927, the year after NBC was launched, twenty-six-year-old Paley signed a fifty-dollar-a-week advertising contract between his family-owned cigar company and Philadelphia radio station WCAU for "The La Palina Hour." The show was a resounding success, La Palina Cigar sales doubled, and William Paley had begun an extraordinary career in broadcasting.

With the sort of shock-of-recognition shrewdness that was characteristic, Paley—given financial backing by his father—bought a failing chain of sixteen radio stations in 1928, and United Independent Broadcasters was reincarnated as the Columbia Broadcasting System. Within two years CBS—at first facetiously referred to as "Paley's Follies"—had burgeoned to seventy stations and showed a net profit of $2.35 million. The rivalry that quickly sprang up between CBS and NBC in the late 1920s and early 1930s had a vital impact on the industry, for it led directly to a network system that, rather than being a government monopoly as in Great Britain, was private and commercial.

One of the greatest quandaries facing the broadcast industry as the networks took shape—and particularly as the Depression came on—was how to change radio from a technological toy into an everyday necessity. Paley felt that the answer lay in programming, but, in order to have a real effect on what programs his network offered, he first had to restructure CBS's relationship to its affiliates. And rather than imitate NBC's system—in which affiliates were charged for sustaining (unsponsored) programs but reimbursed for sponsored network programs—Paley made CBS's entire sustaining schedule available free to affiliates. The offer was an enormous boon, especially for smaller stations: out of ten or twelve hours a day, an affiliate could use as many programs, or as few, as it chose. In return, Paley had the right to

CBS founder William S. Paley.

Charcoal on paper by René Bouché. William S. Paley

secure any part of the affiliate's schedule for network programming; the network also had the right to hire national sponsors without consulting an affiliate.

Once he had secured regular time-slots for network scheduling, Paley was free to focus on programming, which is where he primarily established his legacy. By the mid-1930s, when the average family listened to the radio for more than five hours a day, vaudeville and comedy were beginning to overtake music in audience popularity. So Paley, who had an instinctive sense of popular taste, made it a point to develop new talent, giving regular airtime to such discoveries as Frank Sinatra, Bing Crosby, and Kate Smith. He also—displaying a wonderful flair for blatant banditry—raided NBC for established stars like Major Bowes, Al Jolson, and Eddie Cantor.

The years from 1936 to World War II were marked by a spectacular surge of creativity, which broadcast historian Erik Barnouw has called a "radio renascence." Part of this surge centered on radio's emerging potential as a "theatre of the mind." The "Columbia Workshop" premiered in 1936 and became a showcase for radio's affinity for drama. This ethereal empathy produced such highlights as Archibald MacLeish's *The Fall of the City*. Norman Corwin—later called the "bard of radio's Golden Age" by Studs Terkel—debuted in "Norman Corwin's Words Without Music"; and there was Orson Welles and the "Mercury Theatre."

But Paley's real interest was in the development of network news. Radio news coverage was minimal in the late 1920s and early 1930s, with CBS offering only a daily five-minute report drawn largely from newspaper headlines. Wary of radio's infringement on their territory, the print media created roadblocks: the American Newspaper Publishers Association in 1931 proposed that radio's use of wire services be carefully regulated; two years later the wire services prohibited sale of their news-gathering service to radio altogether.

Paley joined the fight and established a CBS news service, with bureaus in New York, Washington, D.C., Chicago, and Los Angeles, and with freelance stringers in every city having a population of more than 20,000. Radio audiences responded with enthusiasm, and their eagerness to hear news instantaneously—sometimes with bulletins interrupting regular programs—helped to magnify radio's role in everyday life. By 1935 Paley was putting together the gifted team of broadcast journalists—led by Edward R. Murrow as "director of talks"—who would make CBS the preeminent news network.

The coming of war fueled the rise of broadcast journalism. At the beginning of the *Anschluss* in March 1938—when Hitler seized Austria—Paley saw at once that war in Europe was imminent and that it was vital for Americans to be able to get a full grasp of such a war's implications. Stringing CBS reporters across Europe, he orchestrated a "News Round-Up" of simultaneous broadcasts relayed from London, Berlin, Vienna, Paris, and Rome to New York. The remarkable success of this feat was one that Paley himself termed "probably the best job ever done in radio broadcasting."

As it happened, the late 1930s signaled the onset of radio's finest hour: World War II would be a radio war, with events in Europe and the Pacific instantly accessible at the touch of the dial. But the war was also radio's last stand as the predominant information and entertainment medium, for, when the fighting stopped, war-weary Americans turned increasingly to the flickering small screen of "visual radio."

Paley considered television's future undetermined: there were only six transmitting stations and a few thousand television receivers in 1946. Television networks did not exist, much less any set pattern of programming and production; in fact, an American television system had yet to emerge. Ironically, CBS television would be launched out of profits from the radio network: as Paley admitted, "Radio helped give birth to television and . . . it was the growth of television that radically changed radio. Killing off its popularity."

Once again, as in the formative years of radio, rivalry between the commercial networks determined the outcome of the industry. Television in the late 1940s seemed to be the perfect arena for visual entertainers, and few were more visually graphic than the old vaudevillians. Ed Sullivan brought the dog-and-pony school of entertainment to CBS's "The Toast of the Town" in 1948—the same year Milton Berle dragged his act to high ratings on NBC. But in addition to such tit-for-tat, Paley began the same kind of attack he had used against NBC in the early 1930s. First he gleefully raided that network for its biggest star, Jack Benny. Others followed, and CBS soon boasted a lineup that included George Burns and Gracie Allen, Red Skelton, and Edgar Bergen. At the same time Paley developed such new CBS stars as Lucille Ball. By the early 1950s the once-junior network found itself the leader in both news and entertainment.

Paley's insistence on quality programming, and his self-described "gut instinct" for tapping into the "mystical connection between the broadcaster and his audience," brought him widespread recognition as broadcasting's preeminent figure. In his invention and guidance of CBS, he not only helped to shape what went out over the air, but—through the intense rivalry established between CBS and NBC—quickened the commercial network system that defined American broadcasting.

Frank Stanton born 1908

Soon after Dr. Frank Stanton became president of CBS in 1946, the *New Yorker* twitted him as "one of the few men in the history of business to achieve success despite the handicap of a Ph.D." He earned his degree in industrial psychology at Ohio State in 1935, where his dissertation dealt with devising a scientific method for measuring the mass reaction of radio audiences. In connection with this research, he invented the first automatic recording device to measure home listening habits. He also researched radio advertising and determined that advertising copy was more effective when heard than when seen. To bring this finding to the attention of the radio industry, he perspicaciously sent a copy of his paper to CBS, where vice-president Paul Kesten called the report "good red meat for my meat grinder" and hired Stanton for fifty dollars a week.

By 1938 Stanton was head of the CBS research department. In these same years, he and Dr. Paul F. Lazarsfeld developed the "program analyzer," which could measure audience reaction almost second by second. Hailed as a monumental development, the analyzer pre-tested programs and commercials before small

Frank Stanton and Paul Lazarsfeld with a program analyzer, circa 1939.

CBS Inc.

groups of people. Each person was armed with a pair of buttons that registered either positive or negative reactions, and at the end a graph was produced to show how listeners had responded. By the mid-1940s an improved version of this device—affectionately known as "Big Annie"—was developed to record the response of larger groups. In each instance the analyzer was credited with substantially improving the quality of programming and honing the particular angle of advertising. It is a prime example of Stanton's passion for useful research: as pollster Elmo Roper once said, "Frank knew that research was a doomed duck unless it was used to produce action."

But research was only part of Stanton's pioneering contribution to the development of radio and television. Over the course of his career, he became known as the "statesman of broadcasting" because he consistently argued for giving the electronic communications media the same rights and protections as those afforded the print media. Such protection was "overwhelmingly in the interests of the people."

Stanton's overall contribution to broadcasting was facilitated by his working relationship with William S. Paley. By the beginning of World War II, Stanton was an administrative vice-president in charge of research, sales, building construction, press-agentry, maintenance, and operations. He also supervised several CBS-owned stations and commuted to Washington, D.C., as a wartime consultant to the secretary of war and the Office of War Information. As *TIME* magazine commented in a 1950 cover article, "Stanton's success story makes Horatio Alger seem believable."

From the late 1940s on, he became CBS's chief public spokesman, a role that Paley himself never enjoyed. In the first decade of his presidency there—years that saw the rise of television—CBS became not only what has been called "the Tiffany's" of American broadcasting, but a megacorporation. Through it all, Stanton served as the virtual conscience of the industry, defending television's broad appeal as a mass medium for "cultural democracy" and calling for broadcasters to provide responsible standards of scale and balance.

Certainly one of the highlights of Stanton's career in television's formative years was his successful 1960 campaign to get Congress to temporarily suspend the Federal Communications Commission's "equal time" rule, thereby making the Kennedy-Nixon debates possible. It was an accomplishment consistent with his belief that broadcasting has an obligation to give the American people full access to information.

Perhaps his greatest legacy, as *Forbes* magazine has noted, will be his role as "the most literate, tireless, and influential spokesman for the broadcasting industry."

Frank Stanton

Tempera on artist board by
Boris Artzybasheff, 1950.
National Portrait Gallery,
Smithsonian Institution; gift of
Frank Stanton

Allen B. Dumont with a cathode-ray tube, his invention that facilitated the development of commercial television.

National Museum of American History, Smithsonian Institution

Allen B. Dumont 1901–1965

A pioneer in television electronics, Allen B. Dumont was best known for innovations in cathode-ray-tube technology. After training in electrical engineering at Rensselaer Polytechnic Institute, he began his career in the Westinghouse development laboratory, where he became the engineer in charge of production. In 1928 he joined the De Forest Radio Company and soon devoted his research to television. While at De Forest, he helped build the first television transmitters for the simultaneous broadcast of sight and sound, but was dissatisfied with the use of a mechanical scanning disc that had a maximum picture definition of only sixty lines—10 percent of the definition Dumont thought possible.

Dumont left the De Forest Company in 1931 to develop a workable and inexpensive cathode-ray tube that would make all-electronic television feasible. With an initial investment of one thousand dollars he succeeded, and it went into commercial production. Yet the demand for television technology was slow to emerge in the mid-1930s; most people had never heard of television, much less considered putting one in their living room. So Dumont temporarily turned back to radio research and invented the "magic eye" cathode-ray radio-tuning indicator. This he sold to RCA, and with these profits he was able to again turn to the development of television.

Dumont was convinced that it would eventually be possible to build a national television system that would be as widespread as radio networks were in the mid-1930s. In 1935 the Allen B. Dumont Laboratories began the design and production of all-electronic television receivers; two years later the first were manufactured, and by 1938 the first television receivers were marketed in the United States. When RCA inaugurated scheduled telecasting at the 1939 New York World's Fair, Dumont televisions were the only receivers readily available for commercial sale.

World War II interrupted the development of the television industry, with most electronics research being devoted to war production. Dumont Laboratories regeared for defense research and made important contributions in the manufacture of radar components. But the company also did continue its development of television, with employees broadcasting at night from Dumont's New York station—one of only three commercial television stations in the country to continue to broadcast during the war.

With the end of the war, Dumont returned immediately to television production. On May 19, 1945, the Dumont network broadcast from New York through Philadelphia to an experimental station in Washington, D.C., W3XWT. But Dumont was unable to compete against the established giants. By 1955 the network had reverted to being a small number of local stations.

One of the most ingenious and visionary of the television pioneers, Dumont himself once observed that his essential mistake was not in the direction of his research—for his inventions had helped make modern television broadcasting possible—but that he had put his discoveries on the market a decade too soon.

A 1939 Dumont television.

National Museum of American History, Smithsonian Institution

George Washington Hill, as president of the American Tobacco Company, was a leading exponent of early radio advertising. His most notable campaign was his successful scheme in the late 1920s to promote cigarette smoking for women as a diet aid: "Reach for a Lucky Instead of a Sweet."

Photograph by Richard Carver Wood for *Fortune* magazine, 1936

George Washington Hill
1884–1946

Broadcasting magazine once pronounced George Washington Hill "the single dominant figure in the development of broadcast advertising." Until 1924 advertisers could not even purchase radio time. Through the mid-1920s advertising agencies had little grasp of the listening audience, nor ideas about how to reach it. There was also substantial opposition to the notion of "sullying the ether." David Sarnoff at first was appalled at outcroppings of commercialism, though he soon came to accept—as did William Paley—that advertising was radio's only feasible means of support. By the late 1920s, advertisers found that they could best win brand recognition by naming a radio show after their product. The airwaves soon reverberated with the likes of the A & P Gypsies, the Clicquot Club Eskimos, the Happiness Boys, and a program sponsored by Palmolive Soap that featured the duo of Paul Oliver and Olive Palmer.

The year 1928 was pivotal for the growth of commercial radio, as advertising agencies began to approach radio programming with zeal. One pundit said, "Advertising in this jazz age must hit us as we run," and radio was perceived as providing an instant jolt.

The chief provocateur in radio's acquiescence to advertisers was George Washington Hill, the controversial president of the American Tobacco Company who decided in 1928 to use radio as the chief medium for converting American women to cigarette smoking. "Reach For a Lucky Instead of a Sweet" was Hill's ingenious plan for keeping women thin; that is, smoking became a diet aid. When the candy industry erupted at this anti-sweets plot, Hill backed off slightly, reducing the slogan to "Avoid over-indulgence if you would maintain that modern, ever-youthful figure. Reach for a Lucky instead."

Hill said that he had conceived this campaign one day while driving home: *A very fat woman was standing on the corner, chewing with evident relish on what may have been a pickle but which I thought was a sweet. Then I saw a flapper, who took a cigarette out of her bag and lighted it. I thought how much better it would be if the fat woman had smoked cigarettes instead of eating candy.* Hill happily increased his sponsorship of NBC's "Lucky Strike Dance Hour" for the next year, noting that the radio program had inspired a 47 percent increase in Lucky Strike sales.

Another Hill contribution to radio advertising history came with his Cremo cigar campaign: "Spit is a Horrid Word." Here the implication was that Cremo was the only American cigar manufacturer to use machines to seal cigars—and that all the other companies employed factory workers to "spit-tip" them. Hill also dreamed up "Lucky Strike Has Gone to War" and, in 1942, "LS/MFT," or "Lucky Strike Means Fine Tobacco: so round, so firm, so fully packed, so free and easy on the draw."

Hill's sponsorships included "Your Hit Parade," "Information Please," and the Jack Benny show. He also paid two auctioneers $25,000 a year each to open and close his programs with "Sold American!" Often accused of vulgarity and "circus advertising," Hill was said to be the model for the Evan Llewelyn Evans figure in Frederic Wakeman's 1946 novel, *The Hucksters*. Evans was an "advertising and radio genius . . . who had built and broken more stars than anyone else in radio. . . . He was certainly the General MacArthur of the ad game . . . a showman."

Whatever his indulgence in "super-advertising" hokum, Hill built up the sales of American Tobacco even during the Depression; in the first six months of 1930, Americans bought 100 percent more of his tobacco than in 1929. To a generation that embraced advertising as a veritable theology, George Washington Hill was a prophet.

High Culture

America's initial response to the emergence of radio in the early 1920s was to herald it not only as an agent of democracy, but as a redemptive cultural force. A 1922 article in *Collier's* magazine described radio as a "tremendous civilizer," spreading entertainment "into once dreary homes," reducing the isolation of the hinterlands, and leveling class distinctions. Broadcasting veritably "radiated culture," as one journalist wrote in 1924, and would transform every home into an extension of Carnegie Hall or Harvard University. Another rejoiced, "The day of universal culture has dawned at last." Anyone with a radio could tune in "sermons, speeches, weather, stock market and crop reports, the general news of the day, violin solos. . . ."

Broadcasting's ability to serve as a national unifier was a direct outgrowth of its cultural function. According to one observer in the 1920s, the telegraph and the telephone had woven the country together to a degree, though the nation remained "a coarse fabric with wide meshes." With radio came promise: "How fine is the texture of the web that radio is even now spinning! It is achieving the task of making us feel together, think together, live together."

While functioning as a mass commercial enterprise, broadcasting took its civilizing mission seriously. From the first, radio and television adopted an uncompromising approach to high-culture programming, often providing sustaining (unsponsored) air time to accommodate it. The idea was that access to the arts would create an audience for cultural programming—and it did. A 1940 survey conducted by WGN in Chicago, for example, indicated that 40 percent of its listeners favored more operatic programming.

In years when the networks had a serious commitment to cultural programming, radio and television emerged as major artistic forums. Today, when such programming is confined to specialized FM radio stations and PBS television, it seems remarkable that commercial broadcasting once saw its role as being so all-inclusive.

Agnes de Mille during the "Omnibus" show "Art of Ballet."

Photograph © by Roy Stevens, February 26, 1956

Paul Whiteman 1891–1967

Paul Whiteman, "The King of Jazz," was the foremost popularizer of this revolutionary new music in the 1920s and 1930s. Probably best known for commissioning George Gershwin's *Rhapsody in Blue* in 1924, "Pops" Whiteman helped develop an entire school of musical composition in the jazz idiom and gave it a serious audience.

Whiteman began as a classical musician but learned about both pop music and radio in the navy during the First World War. One night in 1917, while eating dinner in a restaurant in San Francisco's Barbary Coast, he heard an outlaw quintet—clarinet, trombone, cornet, piano, and trap drum—playing jazz. All around him people were loving it. Jazz seemed to be the means, he later recalled, for "making a lot of noise and cutting capers. Jazz gives them the exhaust valve they need."

Whiteman began to think about transferring this sound to an orchestra, and in 1919 he formed a fifteen-piece group that featured muted horns and saxophones—one of the first legitimate chances that instrument had ever been given. With this first dance orchestra, Whiteman introduced "symphonic jazz." Though he scandalized some critics by jazzing up the classics, he won a large following, initially among the Hollywood film colony and then extending to fans up and down the East Coast. The orchestra's Victor recording of "Whispering" sold nearly two million discs. In 1920 they were invited to play at Broadway's Palais Royal nightclub—the occasion marking, it has been said, the inception of the Jazz Age. Four years later, when Whiteman arranged the Aeolian Hall jazz concert in which Gershwin played his new *Rhapsody in Blue*, concert jazz had arrived.

Beginning in 1929 on the weekly "Old Gold—Paul Whiteman Hour," Whiteman became a familiar radio personality, enlisting that medium in his campaign to popularize jazz. The Rhythm Boys, featuring Bing Crosby, helped the show's popularity among collegians and flappers, with arrangements of such current songs as "I'll Get By," "Louise," and "Sweet Georgia Brown." The Old Gold sponsors were delighted to win over this younger group, already targeted as the newest market for cigarettes.

In 1933–1934 Whiteman's colorful arrangements made the "Kraft Music Hall" (with Al Jolson) a radio favorite; his 1937 series, "Chesterfield Presents," emphasized swing performed by Jack and Charlie Teagarden, Frank Trumbauer, and The Modernaires. In 1943 he originated the "Philco Radio Hall of Fame," and in 1947 he began the first coast-to-coast network disc-jockey show. Whiteman broke into television with "Paul Whiteman's Goodyear Revue," "Paul Whiteman's TV Teen Club," and "America's Greatest Bands."

The looming presence of "Pops" helped set the tempo of American music for four decades. Among his alumni were George Gershwin, Bing Crosby, Dinah Shore, Mildred Bailey, and the Dorseys. Whiteman helped earn artistic and critical recognition for jazz by orchestrating this essentially black idiom for a mainstream audience. With the help of his arrangers Bill Challis and Ferde Grofé, Whiteman refined and popularized what he called "the folk music of the machine age."

Paul Whiteman popularized a mainstream commercial version of jazz in the 1920s, helping to win for this black-rooted American music a wide following through his radio broadcasts and recordings.

Broadcast Pioneers Library

Milton Cross 1897–1975

"The house lights are being dimmed. In a moment the great gold curtain will go up." Milton Cross was the voice of the Metropolitan Opera broadcasts—except for Deems Taylor's brief tenure—from their inauguration on Christmas Day in 1931. A tenor himself, Cross began as a ballad singer with WJZ, Westinghouse's Newark station, in 1922, when it was nothing more than a cloister off of a factory restroom.

For decades announcer Milton Cross was the voice of the Metropolitan Opera's Saturday matinee broadcasts.

Opera News

Cross's first opera broadcasts were for the Chicago Lyric, where he performed deftly enough for the Met to choose him as its announcer. The Met broadcasts began with *Hansel and Gretel* on December 25, 1931, although there had been such earlier attempts as Lee De Forest's 1910 experiments in broadcasting *Cavalleria Rusticana* and *Pagliacci*, with Enrico Caruso and Emmy Destinn. *Hansel and Gretel*, though, was the first nonexperimental radio broadcast from the stage of the opera house, and one music historian observed that it was as if "an aural window had been opened."

Live broadcasts of the world's greatest singers were now available across the nation, rather than just to those sitting in the opera house. It was precisely this kind of democratization of culture—or at least accessibility to culture—that radio pioneers such as David Sarnoff had advocated in the early years of broadcasting: one in which the radio listener, whether on a Nebraska farm or in a Fifth Avenue mansion, had equal access to plays, music, sports, or news at the touch of the dial.

Every Saturday at 2:00 P.M. from December until April, for forty-four seasons and more than eight hundred performances, Milton Cross announced, "Texaco presents the Metropolitan Opera." His presentation was genial and straightforward, and he knew his audience: "The majority of our listeners by far are plain, average, American 'folks,' who have never attended an operatic performance (and possibly never will)—but who have made acquaintance with it and come to love it entirely through radio."

Walter Damrosch 1862–1950

When NBC and CBS were organizing themselves into networks in the late 1920s, there was sizable sentiment for using high-culture programming to sell radio receivers. NBC's advisory council in 1927 reported itself "dazed over the vast possibilities of radio as an instrument of education"—and as a selling point for RCA sets. To spread the cultural net far and wide, NBC hired Walter Damrosch, conductor of the New York Symphony, to preside over a "Music Appreciation Hour," which would be beamed to 125,000 schoolrooms. Damrosch was to join such other cultural programming as the "Farm and Home Hour," originating out of Chicago for the heartland farmers; politicians given free air time to debate issues; performers like opera stars Ernestine Schumann-Heink, Maria Jeritza, and Rosa Ponselle; and a plethora of plays, concerts, operas, and lectures—all of which paraded before the NBC microphones in 1927 and reached a prime-time audience of fifteen million. "The Music Appreciation Hour" also marked an important turning point in network programming: NBC's use of noncommercial, or sustaining, airtime for public service programs proved so successful that others soon flooded the air. The culture push seemed to pay off, for in 1928 money was spent at a faster rate on radio sets and components than on cars.

Light entertainment soon took over, though, as audiences turned to serials such as "Amos 'n' Andy," to variety performers like Rudy Vallee, and to comics like Eddie Cantor. As William Paley of CBS said, "Those who put on the most appealing shows won the widest audiences, which in turn attracted the most advertisers, and that led to the greatest revenues, profits, and success." Yet high culture was always retained as an enclave of legitimacy on the mass medium, and it did indeed gather in a mass audience that would otherwise never have listened to Shakespeare or Verdi or Beethoven.

Walter Damrosch brought impeccable European and American symphonic and operatic credentials to radio. In October 1923 on WEAF, he gave the first lecture-recital to be broadcast. For Damrosch, radio meant bringing classical music to millions, rather than to mere thousands in a concert hall. He also believed that America would never truly nurture a musical culture until music education was begun in the schools. When Damrosch's "Music Appreciation Course" was launched on NBC, it was immediately apparent that he had struck the right chord, for the course was an overwhelming success. It was estimated that six million American schoolchildren a year listened to him conduct and explain classical music, and his series was hooked up eventually to Mexico, the Bahamas, Bermuda, Cuba, and Latin America. A likable, warm performer, Damrosch grew to be "the best-known and best-loved musician in America." He was fondly recognized as "the Nation's music teacher" until his series ended during World War II, when he was eighty years old.

Walter Damrosch

India ink and wash on paper by Miguel Covarrubias, circa 1937. Harry Ransom Humanities Research Center, The Iconography Department, Nickolas Muray Collection, The University of Texas at Austin

Alexander Woollcott 1887–1943

Radio's "Town Crier" regularly clanged his show to order, shouting, "Hear ye, hear ye. . . . This is Woollcott speaking," and carried his audience off to a Cloud-cuckooland of gossip and literature. Alexander Woollcott—author, drama critic, inspiration for the Kaufman-Hart play *The Man Who Came to Dinner* (1939)—brought his highly individualistic style to radio in 1929 on WOR; switching to CBS the following year, he established himself as "The Town Crier" in 1933. The flamboyant Woollcott, once the centerpiece of the Algonquin Round Table, was an unusual personality for the public to embrace—even in years when individuality was indulged. He was acerbic, maudlin, funny, smart; he presented the news "as I note it in the passing crowd" and proposed "to talk of people I've seen, plays I've attended, books I've just read, jokes I've just heard." The result was a discursive patter of cultural allusions, with Woollcott's monologues occasionally interrupted by music.

The real substance of the program was Alexander Woollcott himself, not his "news." Eccentric in appearance and behavior, he constructed himself as a living legend, an animated stylist. He was the apotheosis of the Personality, and radio reveled in such radiance. Woollcott was urbanity personified—the sophisticated New Yorker. And through his conversational, anecdotal manner of broadcasting, he made his listeners feel as if they, too, were sharing a select cultural experience.

Woollcott did not see himself so much as a "critic" as a middleman—much as Clifton Fadiman, erudite host of "Information, Please," once called himself "a kind of pitchman-professor, selling ideas, often other men's, at marked-down figures, which are easier to pay than the full price of complete intellectual concentration." Mediated by interlocutors such as Woollcott—and Fadiman, and Damrosch, and Toscanini—the mass medium created a consumable culture for its audience, to do with as it chose.

For the Depression audience of the 1930s, Woollcott played the panoply of human emotions. His stories had a point, whether to offer encouragement, or nostalgia, or kindness. Sentimental pap? Sometimes, but as critic John Mason Brown once said, he "can persuade you there are truffles on his tonsils."

On January 23, 1943, Woollcott was appearing on a CBS "People's Platform" broadcast when he was stricken by a heart attack. He collapsed at the mike and died about four hours later.

Critic and wit Alexander Woollcott, radio's resident "Town Crier."

Pastel on canvas by Leonebel Jacobs, circa 1937. Hamilton College; gift of the Estate of Alfred Lunt and Lynn Fontanne

Arturo Toscanini 1867–1957

At the age of seventy, Arturo Toscanini embarked on the phase of his career that would create his most permanent legacy—as conductor of the NBC Symphony. He had been musical director of La Scala before coming to New York in 1908 and firmly establishing the Metropolitan Opera's international reputation. He returned to Italy in 1915, and then came back to America to be principal conductor of the New York Philharmonic from 1930 to 1936. At the zenith of his fame in the 1930s, he also conducted historic performances at the Bayreuth Festival, and in Salzburg and Vienna, and inaugural concerts in Palestine of

Arturo Toscanini conducting the second movement of La Mer *at Carnegie Hall, March 4, 1947.*

Photograph © by Robert Hupka, from *This Was Toscanini* by Samuel Antek and Robert Hupka, Vanguard Press, 1963

what would become the Israel Philharmonic Orchestra.

No other conductor could have better suited the vision of David Sarnoff, who since his 1916 "Music Box" memo had strongly advocated the use of radio for cultural uplift. In 1927 Sarnoff asked, "Through what other medium of communications could a musical, cultural, and entertainment service have been rendered to many millions of homes throughout the world?" In the mid-1930s at CBS, William Paley had guided radio's emergence as the prime arena for drama and poetry through such programs as the "CBS Workshop," and "Norman Corwin's Words Without Music." David Sarnoff, seeing the cultural gauntlet being thrown, responded by campaigning to get the world's foremost conductor to preside over NBC's own symphony. It would be a sensational coup.

The problem was to convince Toscanini of this great opportunity. Sarnoff induced Samuel Chotzinoff, music critic of the *New York Post* and a close friend of the maestro's, to persuade Toscanini to leave Milan for Radio City. Rebuffed by cable, Chotzinoff went over to speak with Toscanini directly. The interview began with Chotzinoff asking him if he knew what "NBC" was: "No," came the reply. When the emissary explained what the network was, and the huge audience it generated, Toscanini agreed. Chotzinoff felt that he had achieved "the greatest scoop of the century."

The hand-picked NBC Symphony of ninety-two artists was assembled by Artur Rodzinski of the Cleveland Symphony, and Studio 8-H was acoustically renovated and enlarged to seat an audience of 1,400. The inaugural concert took place on Christmas night 1937, beginning a legendary tradition that would last until 1954. Early wireless inventor Lee De Forest wrote to Sarnoff that the NBC Symphony was "the capstone to the structure of broadcasting, the realized perfection of my life's dream." And, though huckster George Washington Hill, NBC's biggest sponsor (of Lucky Strike and Cremo fame), told Sarnoff that "Symphonic music has no place in a mass medium," *Fortune* magazine did an analysis in January 1938 which showed that "of all the people in the United States—Negroes, poor whites, farmers, clerks, and millionaires—39.9 percent have heard of the name of Arturo Toscanini; and of those who have heard of him, no less than 71 percent can identify him *as an orchestral conductor.*" *Fortune* concluded that the Toscanini audience was "certainly a mass audience."

Toscanini remained in America during the war years and afterward kept returning for "just one more year." From 1948 to 1954 the concerts were simulcast on radio and television. His television debut in 1948—at the age of eighty-one—was, according to *Newsweek*, "terrific." Chotzinoff directed the three cameras to concentrate first on Toscanini, second on the orchestra, and third on the first-desk players or sections featured at the moment. Seeing Toscanini from the other side of the podium was a revelation, watching him sing along, scream, and otherwise exhort while he conducted. In one startling moment during "The Ride of the Valkyries," *Newsweek* reported that Toscanini, "waxed mustache bristling and nearsighted eyes burning, looked like an inspired Satan presiding over all the fires in hell."

Toscanini's last performance was on April 4, 1954. He ended the program with the Prelude to *Die Meistersinger* and, as *New York Times* music critic Olin Downes wrote, "With the last chord, the master hand which has so long and gloriously held the baton, dropped it to the floor, as Toscanini, almost before the sound of it had stopped, stepped from the podium, left the platform and never returned to the stage."

A few disgruntled critics, then and now, have bewailed Toscanini's decision to spend seventeen years plying culture to the masses. One has recently blasted the maestro for putting himself "in the same sphere as the 'music appreciation' vogue, Goodhousekeeping and Reader's Digest." Toscanini never worried about that himself, even when Chotzinoff had first asked him to conduct the NBC Symphony in 1937. He had simply said, "I'll do this."

Leonard Bernstein born 1918

Leonard Bernstein's television career began with the November 14, 1954, broadcast of "Omnibus." Bernstein recalled getting involved in the program—and realizing "the tremendous power of the medium, the power it could have in terms of music"—through a producer of his 1944 musical *On the Town*, who had come to work with Robert Saudek to produce "Omnibus." They had been planning a segment on Beethoven's Fifth Symphony, focusing on the composer's sketchbooks.

Bernstein took on the assignment, developing a program based on the first movement, and discovered that what was important was not just his ideas about Beethoven's creative process, but that he "could share it with millions of people face to face, eye to eye, nose to nose." And he could use visual aids, adding an entirely new dimension. On the program, Alistair Cooke introduced Bernstein, who appeared standing on a stage painted with the entire score of the first page, white notes on black floor. Musicians stood on designated lines, and Bernstein then "dismissed the ones Beethoven had dismissed in his own mind. Because he begins the Fifth Symphony only with strings and clarinet, out went the flutes, oboes, horns, trumpets, leaving only the people who were relevant standing there."

The program's visual impact was stunning, and combined with Bernstein's exuberant and articulate manner, it created a sensation. He returned often to "Omnibus" to do segments on what makes jazz jazz, why an orchestra needs a conductor, what makes Bach Bach, why modern music sounds strange, what musical comedy is, and why opera is grand. The "Omnibus" programs formed the nucleus for Bernstein's own series, "The Young People's Concerts." With these Emmy-winning programs, Bernstein not only attracted a vast television audience but held its attention. Television proved an ideal medium for him to transmit what he termed "the joy of music," and that is just what he did: he translated its intricacies and made it fun.

Leonard Bernstein

Oil on canvas by René Bouché, 1960. Leonard Bernstein

Sylvester Weaver born 1908

As television sought its form after World War II, Sylvester "Pat" Weaver was one of the few to recognize the young medium's unique potential to be more than visual radio. In a 1952 internal NBC memo called "Opportunity," Weaver outlined the quandary facing even good television: "[It] is an art form on its own . . . [but] Berle works from a vaudeville house . . . Godfrey works a radio show with the camera on, [Max] Liebman [producer of "Your Show of Shows"] presents a Broadway show. . . . The best television, except for its communications use, has been the radio-with-pictures formula." As vice-president and then as president of NBC, Weaver spearheaded the drive to raise the camera's eye above the middlebrow, arguing in "Operation Frontal Lobe" for the need to uplift the entire level of television programming. Even though, as a mass medium, commercial television was obliged to program so that the largest audience would tune in, Weaver argued that the networks also had a responsibility to enlighten.

Pat Weaver with J. Fred Muggs.

Photograph © by Philippe Halsman. Lee Lawrence

Weaver's grand design was to use entertainment to get people to watch television, and then to "inform them, enrich them." In practice, this meant that evening programming began with comedy at 8:00 P.M. to pull in the audience and then proceeded to something more substantive, such as the "Philco Hour." In Weaver's tenure, NBC had as many as eleven live hour dramas on each week.

Known as NBC's "thinker-in-chief," Weaver worked for the Young and Rubicam advertising agency and the American Tobacco Company in the 1930s, helping to plan and produce radio shows. In the last year of the war, he produced the "Command Performance" radio show for overseas troops. When he joined NBC in 1949, there were 2 million television sets in American households. He decided that good programming would sell more sets; when he left NBC in 1955 there were 33 million sets in America. His view of television stressed its revolutionary potential: "For the first time, the average man [finds] himself a participant in the world of his own time." He feared that it "would become just a toy if there wasn't anyone to whip it into shape." And he saw himself as just "the boy to do the whipping." Rather than a "living-room toy," television should become "the shining center of the home."

One of the first shows he was responsible for selling to the network was "Your Show of Shows," starring Sid Caesar and Imogene Coca. In 1950 "The Colgate Comedy Hour" was inaugurated, rotating the biggest names in comedy and offering viewers a wide spectrum in variety entertainment. Eddie Cantor, Bob Hope, and Abbott and Costello all made their television debuts on the "Comedy Hour." Weaver also encouraged producer Fred Coe to develop "The Philco Playhouse," thereby launching an era in which live television drama flourished.

Weaver invented two uniquely television genres. The first proved a bonanza for sponsoring a fringe area of TV—early-morning and late-night programming. The "Today" show, hosted by Dave Garroway, premiered to media catcalls at Weaver's claim that the program would

Scene from one of Pat Weaver's specials, Peter Pan, *starring Mary Martin.*
NBC

revamp the nation's early-morning viewing habits; *Billboard* headlined "Weaver's Folly" at the idea that people would watch television as they brushed their teeth, put on their pants, or ate their oatmeal. By 1954 the "Today" show was NBC's most profitable program. The "Tonight" show—first with Steve Allen, then with Jack Paar—won an insomniac audience with its celebrity interviews and humor. And, on Sunday afternoons, Weaver introduced "Wide, Wide World," the first live coast-to-coast look at everyday American life.

Weaver's second major contribution to programming was the "spectacular"—the live, prime-time, high-budget ninety-minute special that preempted regular shows and was intended to give America something to talk about the next morning (while watching the "Today" show). Weaver's purpose was to give TV programming a creative flexibility and to refresh viewer habits occasionally. Mary Martin's live performance as Peter Pan most notably fulfilled Weaver's design and enthralled the nation.

Weaver also orchestrated NBC's introduction of the compatible RCA color television system (able to receive both color and black-and-white transmissions); the inaugural colorcast was Menotti's *Amahl and the Night Visitors* in 1953. And, in a major innovation in the relationship between sponsors and programming, he instituted the "magazine concept" to NBC: the idea was that the network rather than sponsors would henceforth control the programs and sell time on them—as magazines sold space to advertisers. In practice, this led to multiple sponsorship of single programs, instead of one advertiser buying the entire show—thereby allowing the networks to take full charge of program content.

Weaver's inspired unorthodoxy had a vibrant impact on television in its adolescence. He called himself an "information optimist"—someone who saw his responsibility as exposing people "to things in which they have expressed no interest, but . . . in which they would have expressed interest if they *had* been exposed to them."

Robert Saudek born 1911

In 1951 the Ford Foundation set up a radio-TV workshop under the direction of Robert Saudek, then ABC's vice-president for public affairs. Saudek wanted to develop a quality network series that would attract a mass audience large enough to interest commercial sponsors. As he described it then, the series was to provide entertainment and enlightenment for "the largest possible 'middle-brow' American audience."

The result was "Omnibus," a program that owed its title to a Saudek dinner conversation about a show that would be "kind of an omnibus of ideas." He envisioned the series as a clearinghouse of loosely constructed entertainments, "which might stimulate the curiosity of a television audience in fields of human achievement which many people had always considered to be unexciting." *TV Guide*, somewhat less sanguinely, called this cultural catchall television's "biggest gamble."

The ninety-minute mélange of the arts, sciences, history, drama, and politics known as "Omnibus" debuted on November 9, 1952. Saudek chose as the program's host British-born journalist Alistair Cooke, the chief American correspondent for the *Manchester Guardian*, who had just won a Peabody Award for his exceptional BBC radio program, "Letter from America." Cooke had combined his interests in theater, film, journalism, and American culture for two decades before coming to "Omnibus." He soon established himself on the small screen as both a masterful host-interlocutor and the very embodiment of quality television.

Saudek had created "Omnibus" "in the belief that the will to advance television as an art and as a vehicle of advertising exists among broadcasters, sponsors, creative talent and viewers." In 166 "Omnibus" programs he proved just that. Among the series' best-remembered shows were Leonard Bernstein creating the "interiors" of Beethoven's Fifth Symphony, the debut of Mike Nichols and Elaine May, Orson Welles playing King Lear, and Frank Lloyd Wright dis-

Robert Saudek with Joseph Welch preparing for an "Omnibus" show about the United States Constitution.

Photograph © by Roy Stevens, October 17, 1957

Alistair Cooke, the host of "Omnibus."

CBS Inc.

cussing architecture. Others who appeared on the Sunday afternoon series included Leopold Stokowski, Agnes de Mille, Benny Goodman, Jacques Cousteau, Peter Ustinov, Grandma Moses, William Faulkner, and James Agee. Sponsored by the Ford Foundation for five years and then carried on independently by the Robert Saudek Associates, "Omnibus" was a milestone of television's Golden Age.

Comedy

In its first decade, radio developed the kinds of programs and personalities that defined broadcasting's future. Music initially played an important role, dominating two-thirds of all airtime in the mid-1920s, but by the early 1930s comedy ruled the air. The Depression had caused the collapse of vaudeville, and performers flocked to radio stations as they once had to the stage of the Palace Theater. As it happened, the vaudevillians arrived just in time to rescue radio itself: in 1930, three-quarters of those who owned radios tuned in at least once a day. Three years later, that figure had fallen to just over half—a remarkable drop in listenership that sent networks and advertising agencies into a frenzied search for new talent. The search succeeded, and a fresh crop of comedians soon sent ratings soaring. Families gathered around their radios every night to hear Will Rogers, Jimmy Durante, Burns and Allen, and Jack Benny.

Comedians also fueled the popularity of television between 1948 and the mid-1950s. Many of radio's biggest stars—though not all—made the successful transition to the small screen, and a new generation of comedians also appeared. Comedy on television succeeded as it had on radio, and for the same reason that *Radio Guide* had reported in 1936: it was "Main Street entertainment, pure and simple."

George Burns and Gracie Allen

Photograph by Thomas Harrison Evans. Harvard Theatre Collection, Harvard University

*"The Happiness Boys"—
Billy Jones and Ernie
Hare—with Helen Hann,
WEAF accompanist.*

NBC

"The Happiness Boys"
Billy Jones 1889–1940
Ernie Hare circa 1884–1939

Billy Jones and Ernie Hare formed radio's first comedy team in 1921. Two years later, while radio was still uncertain about who would pay its bills, Jones and Hare began a program sponsored by Happiness Candy. "The Happiness Boys" broadcast over New York's WEAF, with tenor Jones and baritone Hare beginning each program with their theme, "How do you do, everybody, how do you do," followed by a half hour of "song and patter" gleaned from years of vaudeville.

In early radio, when amateurs dominated the airwaves, Jones and Hare were *sui generis*: not only were they seasoned stars of vaudeville, but their extensive recording career had given them considerable experience in effectively using a microphone to reach an unseen audience. Along with this sheen of professionalism, Jones and Hare were unique in having a sponsored weekly program. Through the mid-1920s the question of advertising on the air was still under debate, and as late as 1926 most shows on radio were onetime affairs. The idea of a regular, weekly program—one introduced each time by an identifiable commercial jingle—was a departure for radio and, more than that, a beginning.

Jones and Hare—first as the "Happiness Boys" and later, with changes of sponsors, reincarnated as the "Interwoven Pair" and the "Flit Soldiers"—reigned as radio's most popular comedy team for a decade, giving an estimated two thousand broadcasts, singing ten thousand songs, and telling twenty-five thousand jokes. Although they were swept from their pinnacle by the tide of stage stars who flooded the air beginning in 1932, Jones and Hare continued to perform through the 1930s.

Ed Wynn 1886–1966

Philadelphia-born Ed Wynn escaped from his father's millinery business and became a vaudeville headliner by the time he was nineteen. In 1914 he joined the Ziegfeld Follies, and in 1921 he wrote and staged his greatest hit, *The Perfect Fool.*

The Perfect Fool provided the occasion for his first radio appearance. Aired live over WJZ in Newark on February 19, 1922, it was the first stage show ever to be broadcast. But Wynn didn't return to radio for ten years. Vaudeville was still in high gear in the 1920s, and performers looked at radio with great uncertainty, often viewing it as simply a passing fad. Radio also ate up treasured material in seconds and, combined with its notoriously low pay, seemed to glow with a lack of professionalism. By the early 1930s, however, networks began to worry about a drop in evening listening and lured vaudevillians back on the air. It had become obvious that radio was a market begging for heroes: broadcast historian Erik Barnouw has pointed out that, with the Depression, "as theatre and film audiences shrank, home audiences grew. According to social workers, destitute families that had to give up an icebox or furniture or bedding still clung to radio as to a last link with humanity." Since the economy had severely depressed the vaudeville stage, the progression to radio seemed natural; and in fact radio, until the late 1930s, came to depend upon using stage performers rather than developing new talent itself.

The influx of vaudevillians established comedy as king of the air, and Ed Wynn rode near the top of the current. Texaco had wooed him back in 1932, giving him a live audience and five thousand dollars per half-hour show. As the Fire Chief, Wynn insisted on maintaining his essentially visual comedy; he called himself a "method comedian," because he depended on how he presented his material more than on the material itself. The program alternated music with humor and featured Wynn's countless changes of costume and bottomless supply of jokes. A typical shtick was his impersonation of Will Rogers: walking

Ed Wynn on "Texaco Star Theater," January 5, 1936.

CBS Inc.

back and forth across the stage, looking bewildered as only Ed Wynn could look bewildered, and dragging a rope, he looked at the audience and said, "Either I've found a rope or I've lost a horse."

Despite his emphasis on sight gags, Wynn had great success on radio, and his show was one of the most popular on the air during the mid-1930s. Drama critic Joseph Wood Krutch wrote that "No one can exceed him in solid, impenetrable asininity, but no one can, at the same time, be more amiable, well-meaning, and attractive." His popularity lasted until 1937, when radio comedy metamorphosed into situation comedies and the likes of such new personalities as Bob Hope and Fibber McGee and Molly.

Will Rogers 1879–1935

Born in Oologah, Indian Territory (now Oklahoma), Will Rogers once told a Boston audience, "My ancestors didn't come over on the *Mayflower*—they met the boat." He spent his youth cowboying with such troupes as Texas Jack's Wild West Circus, calling himself "The Cherokee Kid" and performing as a rope artist and roughrider. Adding jokes to his lariat tricks, he eventually got involved in vaudeville and began appearing on Broadway in 1912. His reputation was firmly established with the Ziegfeld Follies, in which he starred in 1916–1918, 1922, and 1924–1925. Beginning in 1918 he embarked on a motion-picture career that became increasingly successful with the advent of "talkies."

Rogers and a group of Follies girls broadcast from the *Pittsburgh Post* studio of KDKA in 1922, and he soon began regular radio appearances on such programs as WEAF's "The Eveready Hour"; for one appearance on that program, he was paid an astonishing one thousand dollars.

Ed Wynn as the Fire Chief with foil Graham McNamee at the old NBC Times Square Studio in the 1930s.

Broadcast Pioneers Library

One of Rogers's most prominent roles was as a folksy political philosopher. No public figures—especially Presidents—escaped his good-natured but pointed barbs. He even ran for President himself in 1928, on the "Anti-Bunk" ticket. His slogan was "He chews to run," and his platform promised that "whatever the other fellow don't do, we will." Except for Will Rogers, radio did not serve as a forum for political humor. In fact, fears over possible Federal Communications Commission (FCC) action against stations that allowed political satire largely kept it from the air. But Rogers avoided such censorship: network censors never got the chance to preview his broadcasts because he never used a script.

Will Rogers

From *Great Radio Personalities in Historic Photographs* by Anthony Slide, Vestal Press, 1988

Rogers's homespun humor made him one of radio's favorite stars, but it was not a medium in which he felt totally at ease: "It's made to order for a singer," he said, "and a person making a straight-forward speech, or a talk. . . . But to have to line up there and try to get some laughs, I want to tell you it's the toughest job I ever tackled." The charm of his rambling style was that it truly was extemporaneous, replete with its own rhythm of meandering anecdotes. But radio waited for no one: "They have a time getting me stopped," he would explain on the air, "so I got an alarm clock here, and when it goes off, brother, I quit—even if I'm in the middle of reciting 'Gunga Din' or 'The Declaration of Independence.'"

He insisted on working with a studio audience, which helped—though he still complained about the tyranny of the microphone, of having to speak right into "that radio thing." But the radio audience loved his colorful stories and picaresque language; he seemed a genuinely nice fella, and even his satiric political barbs left no sting. When he died with Wiley Post in an airplane crash near Point Barrow, Alaska, in 1935, Will Rogers was arguably radio's best-loved personality.

America's homespun philosopher, Will Rogers.

Bronze by Jo Davidson, 1935–1938. National Portrait Gallery, Smithsonian Institution

Jimmy Durante 1893–1980

The "Schnozzola," Jimmy Durante.

Watercolor on paper by C. C. Beall. Society of Illustrators Museum of American Illustration

Warm-hearted comedian Jimmy Durante came to radio and television via vaudeville. The "Schnozzola" learned to pound the piano honky-tonk style and to belt out songs in a raspy baritone while performing at Coney Island in the 1910s. In the 1920s he teamed with Lou Clayton and Eddie Jackson to become "the greatest night-club act of all time"; the vaudevillians were also a top draw at the Palace. Durante established himself in musical comedy and wrote dozens of songs whose titles became Broadway catchphrases: "I'm Jimmy, That Well-Dressed Man," "I Can Do Very Well Without Broadway (But Can Broadway Do Without Me?)," and "Did You Ever Have the Feelin' that You Wanted to Go, Still You Have the Feelin' that You Wanted to Stay?"

In 1944 Durante began his biggest radio success when CBS teamed him with Garry Moore. "The Durante-Moore Show" became a wellspring for such Durante-isms as "I've got a million of 'em," "Everybody wants to get into da act," and "I'm mortified!" He also started to use his classic sign-off, "Goodnight, Mrs. Calabash, wherever you are."

In 1950 Durante embarked on a successful television career. The small screen seemed a perfect marriage with Durante: "I like a small place where you know everybody and can kid around." His half-hour Saturday night show became a favorite, but he gave it up in 1956. "That box could be the death of us," he said as he left. "They're going to hate us if we stay on too long." Thereafter, he appeared only once or twice a year as a guest on comedy or variety shows. His lasting popularity was not just the result of his scarcity: as his old partner Lou Clayton once said, "You can warm your hands on this man."

George Burns born 1896
Gracie Allen 1905–1964

The most successful comedy team in broadcasting history, Burns and Allen spanned vaudeville, radio, motion pictures, and television. Both grew up on the boards of old-time vaudeville, meeting in 1922 and forming their own act. At first Gracie was the foil, but George, after realizing that "even her straight lines got laughs," reversed their roles. "I knew right away," he said, "that there was something between the audience and Gracie. They loved her, and so, not being a fool and wanting to smoke cigars for the rest of my life, I gave her the jokes." For the next few years George worked out his role as show business's best "straight man," while Gracie developed her character as the malapropping "dizzy girl."

Married in 1926, they were working at the Palace Theater on a bill that included Eddie Cantor when Cantor asked Allen to be a guest on his radio program. Subsequent Burns and Allen appearances on the Rudy Vallee "Fleischmann Hour" and the Guy Lombardo show landed them their own radio program in 1932 on CBS. George, with his raspy-voiced delivery, and Gracie, with her redoubtable logic, became radio mainstays for the next two decades.

Two of their ongoing stunts attracted particular attention. The first was when Gracie—on their show and on others—conducted a search for her imaginary lost brother. National consternation was such that her real brother had to go into hiding. The second took place in 1940, when Gracie declared herself a presidential candidate for the "Surprise Party." She popped up unannounced on several network programs to answer questions:

Q: *With what party are you affiliated?*
A: *Same old party—George Burns.*
Q: *But don't all candidates affiliate themselves with a party?*
A: *Well, I may take a drink now and then, but I never get affiliated.*

In 1950 Burns and Allen inaugurated a biweekly television series on CBS. Essentially an elaborated version of their

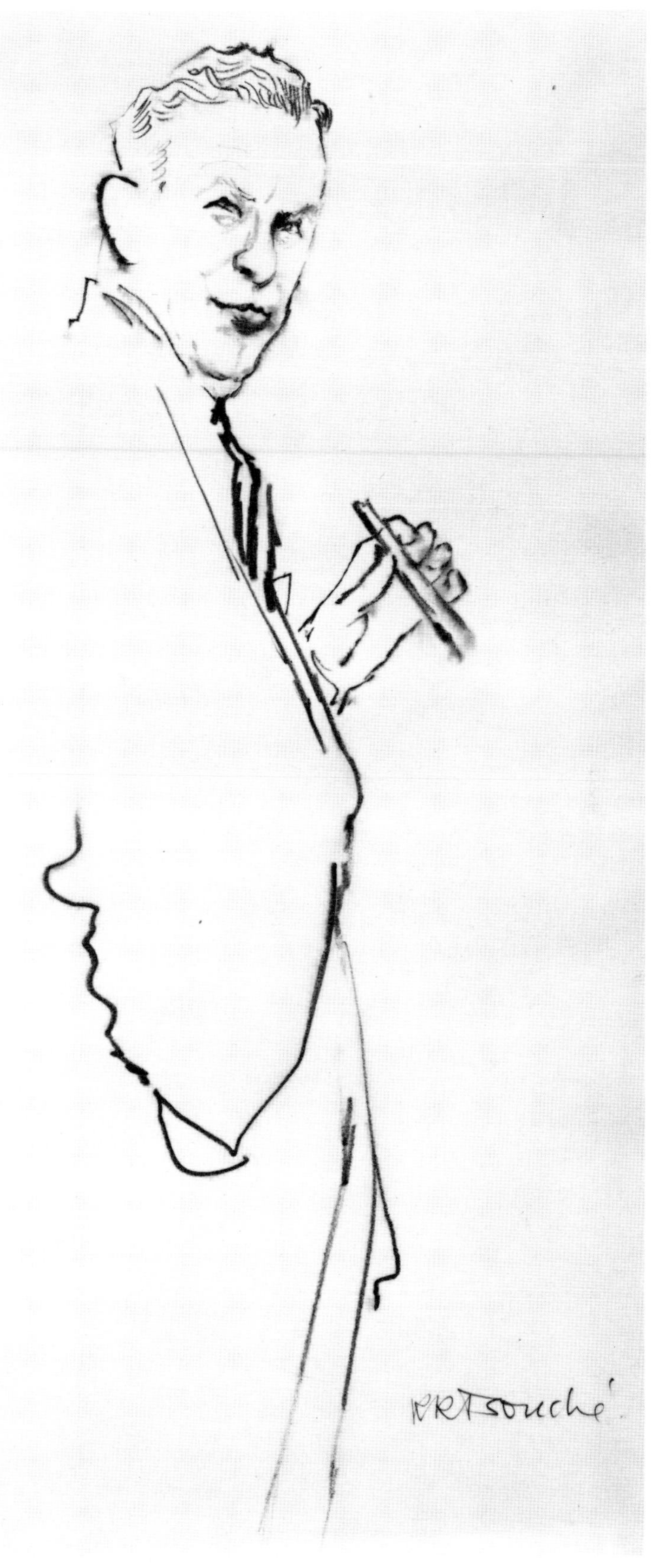

George Burns

Charcoal on paper by René Bouché, circa 1955. CBS Inc.

radio program, they succeeded again by being themselves: they always appeared as George and Gracie. And, while some of radio's biggest comics failed to make the transition to television, Burns and Allen glided over without missing a beat.

George began the show by introducing himself: *For the benefit of those who have never seen me, I am what is known in the business as a straight man. If you don't know what a straight man does, I'll tell you. The comedian gets a laugh. Then I look at the comedian. Then I look at the audience—like this.* [Looks.] *That is known as a pause. And when I'm really rolling, this is one of my ad libs* [surprised look with mouth open].

The show was a combination of situation comedy and vaudeville. George would step in and out of the interior sitcom set to speak directly to the audience throughout the program—a device reminiscent of the narrator in Thornton Wilder's *Our Town*, and one that William Paley is said to have suggested when Burns initially worried about how to begin the show. Regulars included Harry Von Zell, and Harry and Blanche Morton as neighbors, but it was always George who gave the show its focus, whether in or out of character. On a typical episode he sat in his office upstairs watching an all-seeing television. Turning from the set, he faced the camera and explained, "According to my calculations, Harry Von Zell should be over at the Mortons' and by now Gracie should have him mixed up in this too. . . . Let's take a look."

Gracie retired in 1958 and, though the rest of the cast continued for another season, it was never the same; the show went off the air in June 1959. George once said, "Gracie isn't really crazy. . . . She's off-center." She never got accustomed to the camera's glaring red "On-Air" light and in fact failed to notice it for two seasons. One day on the set she gasped, "What's that? I never want to see it again."

What came across most of all with Burns and Allen was their genuine enjoyment of each other. On one episode, Gracie offers hors d'oeuvres to a guest, makes some silly comment, and leaves. The guest leans over to George and says, "The hostess is really a fruitcake." George says, "She's my wife." Guest, thoroughly chagrined, says, "Good hors d'oeuvres," and walks away. George thereupon grins at the camera and says, "Good fruitcake, too."

Say goodnight, Gracie.

Gracie Allen

Charcoal on paper by René Bouché, circa 1955. CBS Inc.

Fred Allen, "the world's worst juggler" and radio's driest wit.

CBS Inc.

Fred Allen 1892–1956

The day after the Fates created radio, they invented Fred Allen. For fun. Giggling. After a night of brut champagne. What they begat was a divine conspiracy between radio and Fred Allen, beginning in 1932 with the (really!) "Linit Bath Club Revue."

Allen had first indulged in show business in high school while working in the stacks of the Boston Public Library, which now, rightfully, houses his vast manuscript collection. He found a book on juggling, became proficient at managing dishes and balls, and traveled the vaudeville circuit from Australia to California as "The World's Worst Juggler" and "Just a Young Fellow Trying to Get Along." In the 1920s he graduated to Broadway revues, appearing in the *Passing Show of 1922*, *The Greenwich Village Follies* (where he met his wife, Portland Hoffa), the first *Little Show*, and *Three's a Crowd*.

His radio career was firmly established in 1934 with his "Town Hall Tonight" program, an hour-long show that was the most literate comedy on the air. The thing he enjoyed most about radio over vaudeville, he once said, was that "the show could not close if there was nobody in the balcony."

But the program for which he was best remembered was the "Texaco Star Theater," where he created "Allen's Alley" in 1942. The denizens of the Alley included Allen as straight man, Kenny Delmar as Senator Claghorn, Parker Fennelly as Titus Moody, Minerva Pious as Mrs. Nussbaum, and Peter Donal as Ajax Cassidy. The toughest problem that Allen had on this show—staged in NBC's cavernous Studio 8-H—was to keep the studio audience laughter restrained enough that the show would end on time; if it didn't, NBC's policy was just to pull the plug.

Allen's feuding with NBC officials—and especially with the network censor—became legendary. On occasion his show was even cut off in mid-program, due to allegedly offensive material. One such censored bit was the following dialogue:

PORTLAND: *Why were you cut off last Sunday?*
ALLEN: *Who knows? The main thing in radio is to come out on time. If people laugh, the program is longer. The thing to do is to get a nice dull half-hour. . . . Then you'll always be right on time, and all of the little emaciated radio executives can dance around their desks in interoffice abandon.*

Allen's devastating wit was also vented in his continuing staged feud with friend Jack Benny; they traded barbs on one another's programs and occasionally invaded each other's studios. Allen cosponsored a contest imploring listeners to submit an ending to the sentence "I Hate Jack Benny Because . . . ," and the resulting avalanche of mail was said to rival Santa Claus's. When it was advertised that the two would face off on Benny's program one night, the demand for audience tickets was so huge that the show had to be broadcast from the ballroom of New York's Hotel Pierre; the listening audience was said to have been second in broadcasting history only to one of Franklin Roosevelt's Fireside Chats. The mark of their long friendship, however, was that on Allen's last radio show in June 1949 his final guest was Jack Benny.

Allen made his television debut in 1950 on the "Colgate Comedy Theatre," but his gloomy countenance and acerbic wit did not translate well to the small screen. In interviews at the time, he strongly lamented the shifting of mass taste away from radio: "On a radio broadcast, anything goes," he told *LIFE*. "You can have a man walking up the side of a building like a spider. You

Fred Allen

India ink on artist board by Al Hirschfeld. The Margo Feiden Galleries, New York City

The Fred Allen radio show in progress.

CBS Inc.

merely plant a suggestion in the listener's mind, and his imagination supplies all the details, all the scenery, props, extras and costumes." He also suggested that "They call television a medium because nothing on it is ever well done."

As a comedy writer and performer, Fred Allen had few peers. Often called a latter-day Mark Twain, he more truthfully stood on his own. As James Thurber said, "You can count on the thumb of one hand the American who is at once a comedian, a humorist, a wit and a satirist, and his name is Fred Allen."

Bergen and McCarthy on the radio.

Broadcast Pioneers Library

Edgar Bergen 1903–1978

For two-and-a-half years in the early 1940s, radio's top-rated program starred a monocled wooden dummy. The "magnificent splinter" was Charlie McCarthy, all thirty-eight inches and forty pounds of him. His sidekick—need we say, his omnipresent sidekick—was Edgar Bergen, a mild-mannered ventriloquist who had invented Charlie while still in high school. The character for the impish dummy was inspired by a tough Irish newsboy; Bergen had the head carved out of white pine by a local carpenter and fashioned the body himself.

After Charlie helped him work his way through Northwestern University, Bergen embarked on the vaudeville circuit, working up from small-time to two-a-day engagements with his top-hatted friend. With the demise of vaudeville in the early 1930s, Bergen and McCarthy—it was always hard not to give both credit—worked up a nightclub routine. One night while performing at the Rainbow Room for an Elsa Maxwell party, Bergen was discovered by Nöel Coward. Word spread of his success, and in December 1936 the duo made their radio debut on Rudy Vallee's show.

Bergen and McCarthy were an instant success—attesting to the listening audience's infinite ability to suspend reality—with Charlie's brashness contrasting sharply with Bergen's shyness. Bergen, for example, would never himself have provoked the risqué dialogue between Charlie and Mae West on one of their shows in 1937; actually, it wasn't the dialogue so much as the suggestive reading of the dialogue that made West's radio debut her last broadcasting appearance for twelve years:

WEST: *You look pretty good to me, Charlie. Come here.*
CHARLIE: *But I thought you only like*

Charlie McCarthy and Edgar Bergen, with supporting cast members Mortimer Snerd and Effie Klinker.

Oil on canvas by Boris Chaliapin, 1944. The Edgar Bergen Family

W. C. Fields, Nelson Eddy, Charlie McCarthy, Edgar Bergen, Don Ameche, Robert Armbruster, and Dorothy Lamour on the Chase and Sanborn "Charlie McCarthy Show."

Broadcast Pioneers Library

tall, dark and handsome guys.
WEST: *Oh, that was my last year's model. This year I'm on a diet. . . .*
CHARLIE: *Oh, Mae, don't, don't, don't be so rough. To me, love is peace and quiet.*
WEST: *That ain't love—that's sleep.*

Bergen called the art of ventriloquism "merely that of cultivating a 'grunt,'" but Charlie wasn't impressed with that. When Bergen explained that ventriloquists were even found among the Eskimos, Charlie retorted, "They weren't found up there—they were chased up there."

Bergen added "Mortimer Snerd" and "Miss Effie Klinker" to the cast, but it was always Charlie who retained the radio—and later television—fans' greatest loyalty. Characterized as "saucy, lethally precocious, and irreverent," Charlie lived up to the legend inscribed across his stationery: "E Pluribus Mow 'Em Down."

Jack Benny 1894–1974

Even now, Jack Benny's carefully created persona, "Jack Benny," conjures up instant images: the perfectly timed pause, the mincing, arm-swinging gait, the exasperated deadpan. His trademark lines became part of the national lexicon: "Hmmm," "Well!," "Now cut that out!" and the classic response to a bandit demanding "your money or your life": silence, then, "I'm thinking, I'm thinking!"

Born in Chicago and raised in Waukegan, Illinois, Jack Benny took his violin to vaudeville when he was eighteen. During one performance, the audience laughed when he told a joke, and America lost a middling violinist. "The sound intoxicated me," he recalled. "That laughter ended my days as a musician, for I never again put the violin back [in the act] . . . except as a gag."

In the 1920s he honed his timing to mastery, reaching the coveted master-of-ceremonies position at the Palace Theater. After a guest shot on Ed Sullivan's radio show in 1932, Benny was soon given his own program on NBC. And with radio he found his enduring niche. By 1937 the show's format was forever set, and Benny's "gang" included Eddie "Rochester" Anderson as the anything-but-servile valet, Irish tenor Dennis Day, announcer Don Wilson, bandleader Phil Harris, and Mary Livingstone, whom he had married in 1927. Benny deftly crafted his own character as "the stingiest man in the world," a regular guy whose self-esteem could be punctured but never totally deflated. One of the ways Benny etched this character so firmly into the popular consciousness was by never wavering from his conceit. He never stepped out of character. Nor, remarkably, did the rest of the cast, many of whom became so subsumed in their roles that they called themselves by their character's names: "Mary Livingstone" was really Sayde Marks, "Dennis Day" was Eugene Denis McNulty, and "Rochester" was Eddie Anderson, but they were ever after known by their stage names.

Benny captured radio's potential as a

The Jack Benny show, November 30, 1948, with Mary Livingstone, Don Wilson, Phil Harris, and Dennis Day.

CBS Inc.

Jack Benny

Photograph by Clarence Sinclair Bull. Alice Becker Levin

mirror to the foibles and frustrations of everyday life and became, as one critic noted, a kind of middle-class counterpart to Chaplin's sympathetic "little fellow." "On the air," Benny said, "I have everybody's faults. All listeners know someone who is a tightwad, show-off or something of that sort. Then in their minds I become a real character." He also thoroughly appreciated radio's power over the imagination. Radio's suspension of visual literalness gave free rein to the realm of suggestion and effect, and Benny masterfully created a whole world that over the years became as recognizable to his listeners as their own: the homey setting, the cast, even the sputtering Maxwell car and the chained bank vault—all were familiar.

Benny revolutionized radio comedy by taking the laughs from the star and handing them over to the cast; he *became* the laugh, whether sawing away on his theme, "Love in Bloom," or sustaining the perpetual joke of being thirty-nine. Once a carnival pitchman promised Benny a quarter if he could guess Benny's correct age; when the carny guessed thirty-nine, Benny was in pure agony over a hopeless choice.

Because his show was more of a short story than a collage of wisecracks, Benny ushered in a new comic age where old vaudevillians could not tread. Eddie Cantor said, "He made all the other comics throw away their joke files." And Fred Allen called his show the model thereafter for comedy structure; Allen also credited Benny with being "the first to realize that the listener is not in a theater with a thousand other people but is in a small circle at the house."

Benny's Sunday-night radio program lasted from 1932 to 1955. Beginning in 1950, he also took on television. Critics say that the television years never matched those on radio, but by the 1950s Benny had become a national icon. Television was in fact a successful medium for him: his instinct for the moment at which comedy clicks surmounted visual literalness, and the double-takes, the silent, nagging pauses, and the frustrated "That's *not* funny!" were classic television.

Fred Allen and Jack Benny, caught in the middle of their infamous "feud."

Fred Allen Collection, Boston Public Library

Jack Benny opening his vault, December 30, 1948.

CBS Inc.

Rochester (Eddie Anderson) starts Benny's Maxwell roadster.

CBS Inc.

Henry Morgan born 1915

In the 1940s radio audiences were accustomed to the plummy tones of Henry Morgan announcing, "Good evening, anybody, here's Morgan." Listeners at this point would either gleefully settle in with their feet propped up, or hustle themselves out of hearing range. The world seemed to divide itself between those who relished Henry Morgan, and those who . . . didn't.

Morgan started as a radio announcer while still a teenager, working his way up to staff announcer at WOR in New York. In 1942 he was assigned a nightly spot called "Here's Morgan," which one critic called "a daily dose of anarchy." The satirist Morgan specialized in a little game called "zap-the-sponsor," as in: "Leading shoemakers report that they scrape more of Wrigley's Spearmint gum off their shoes than any other product." Or, after the Adler Shoe Company introduced their "elevator shoes," asserting they would make the wearer two inches taller, Morgan said, "That claim is correct—you can be two inches taller if you're able to stand up in them." He usually ended his programs with the weather forecast, "Muggy—followed by Tuegy, Wegy, Thurgy, and Frigy."

Norman Corwin called Morgan "the first radiogenic comedian of stature," with a technique "too intimate, too sleight-of-mike for anything but radio." After the war Morgan returned—this time to ABC—and again blithely romped over bigwigs, mocked soap operas, and ridiculed commercials, cheerful in the knowledge that plenty of icons remained to be scorned. The bad boy of radio genially opened his show with "For He's a Jolly Good Fellow," and then proceeded to say something not nice about everyone. He got into a nasty feud with the Eversharp Company, his sponsor in the late 1940s, by changing their slogan for Schick Injector Razors from "Push/pull, click/click," to "Push/pull, nick/nick." When Schick dropped him for "low ratings," Morgan responded, "It's not my show; it's their razor."

Henry Morgan

India ink on artist board by Al Hirschfeld. Henry Morgan

Bob Elliott born 1923
Ray Goulding born 1922

Bob and Ray were hailed by critics as the freshest thing to hit radio after World War II. In their solemn, low-keyed announcer voices—they both had started as serious announcers—they invented radio through the looking glass: radio (and later television) reflected at its most mediocre.

Remembering that "You don't have to smile all the time—this is radio," Bob and Ray adjusted their dial only slightly and tuned in "Mary Backstayge, Noble Wife," "One Feller's Family," and characters like Helen Harkness and reports from the Bob and Ray On-the-Spot News Foundation. It was parody at its sitting-duck best: the apotheosis of the third rate. They perfected the art of "how to seem lusty and purposeful," as Kurt Vonnegut has elegized them, "when less than nothing was going on."

Their first show, "Matinee with Bob and Ray," was broadcast over WHDH in Boston in 1946. Pleased with how nicely the title rhymed, they spent five-and-a-half years doing this program before signing with NBC in New York, where their radio show ran six days a week, from 6:30 to 10:00 A.M. They were instantly a hit, peddling mock commercials, lampooning radio giveaway gimmickry (send today for the Bob and Ray "Home Surgery Kit" or the "Jim Dandy Burglar Kit") and parodying soaps and sitcoms. Every day they featured the continuing story of one of their soap operas, such as "Mary Backstayge, Noble Wife": *the story of a young girl from a deserted mining town out West, who comes to New York to try to find romance and security as the wife of handsome Harry Backstayge, Broadway star, and what it means to be the wife of the idol of a million other women.*

After the soap came ad-libbed commer-

Bob and Ray

NBC

cials for "Make It Yourself" kits: "Why buy an expensive, ready-made car, when from our kit, with parts numbered from 1 to 10,000, you can build a 1927 Jewett for $28.35?" Then the Bob and Ray News Foundation would call on one of its far-flung correspondents, perhaps reporter Wally Ballou at the Bob and Ray Dude Ranch "in secluded New Jersey."

Another high point of the show was the Gourmet Club, where the climactic event was the unwrapping of the Surprise Sandwich. What would it be today? Meat loaf and ketchup on white? Egg salad with bologna slices on rye? With crusts, or without?

By 1951 Bob and Ray had caused enough trouble to be put on television. Their great success was that they looked just as people expected them to look. And they did the same kind of material, caricaturing radio, TV, their sponsors, and other shows. Once they ad-libbed a soap opera parody called "The Life and Loves of Linda Lovely," managed to kill off all the characters in a single episode, and replaced them with the cast of "Mary Backstayge, Noble Wife." Another time they told their listeners to write to the Smithsonian Institution for a copy of the Bob and Ray "Handy Home-Wrecking Kit"; they reportedly got a "stodgy" letter from the Smithsonian, asking them to rescind the offer.

Thank you, Wally Ballou, wherever you are.

Sid Caesar born 1922

Nurtured as the "Admiral Broadway Revue" in 1949, "Your Show of Shows" premiered on February 25, 1950, and for the next four seasons gave viewers a glimpse of what television could be. The basic ingredients were producer-director Max Liebman, a cast consisting of Sid Caesar, Imogene Coca, Carl Reiner, and Howard Morris, and an assemblage of writers that included Caesar, Reiner, Mel Brooks, Neil Simon—and later Woody Allen and Larry Gelbart. The format was a weekly, live, ninety-minute comedy-variety show—that's right: live, ninety minutes, every week. The astonishing thing is not that this brilliantly constructed program achieved cult status, but that these people survived.

Sid Caesar has described "Your Show of Shows" as "a whole new animal" for television. Instead of being an offshoot from radio, it was rooted in theater. Director Max Liebman said that "at the time that television began to capture my attention, I realized that most of the programs being presented really originated in vaudeville and night clubs, or were an extension of radio." What he did was to bring to television an element of Broadway.

The writing had to be done the Wednesday before the Saturday-night program—an hour and a half of routines, sketches featuring regular characters like the Hickenloopers or the rock group "The Haircuts," and one-shot pieces such as "From Here to Obscurity." Caesar worked with the writers six days a week, and the sessions were reportedly intense. One of the writers, Lucille Kallen, said, "Sid boomed . . . Reiner trumpeted, and Brooks, well, Mel imitated everything from a rabbinical student to the white whale of *Moby Dick* thrashing about on the floor with six harpoons sticking in his back. Let's say that gentility was never a noticeable part of our working lives." On Thursdays the show was "put on its feet," Fridays were the technical rehearsals, and Saturdays, the show. On Sundays, Caesar said, he "used to stand under the shower and shake."

Caesar's genius was in capturing a uni-

Sid Caesar

Gouache by Al Hirschfeld. The Margo Feiden Galleries, New York City

Sid Caesar with Imogene Coca

Courtesy Sid Caesar

versal humor in the everyday. "A couple of chairs and a table and you got a sketch. Comedy is not standing up and telling a joke. People are what's funny." What he did was storytelling: "stories that have to be seen and heard from beginning to end, or not at all."

He got his start in show business while in the Coast Guard during World War II, touring with the Coast Guard musical *Tars and Spars* and then appearing in the movie version. The show's civilian director was Max Liebman, who became the godfather for the Caesar television shows, beginning in 1949 with the "Admiral Broadway Revue."

Although Caesar was especially good at such characterizations as the "German Professor," his turns with costar Imogene Coca were standouts. Howard Morris once said, "Her gentle delicacy played against his hugeness—and they would be clanging together against each other." Both were wonderful pantomimists, and favorite targets for their sketches included the ballet, grand opera, and Hollywood musicals. Always they assumed that their audience shared their frame of reference and had enough familiarity with the material—for example, foreign films—to recognize the satire. At the very least, "Your Show of Shows" established that television could offer a lively format for the best of creative comedy. There was a kind of heroism in all that hilarity which has not diminished over time.

Sid Caesar with Nanette Fabray, who costarred as his wife on "Caesar's Hour" from November 8, 1954, to June 18, 1956.

Courtesy Sid Caesar

"The Great One," Jackie Gleason.

Acrylic on masonite by Russell Hoban, 1961. National Portrait Gallery, Smithsonian Institution; gift of Time, Inc.

Jackie Gleason 1916–1987

During the mid-1950s Jackie Gleason was "Mr. Saturday Night." Americans regularly tuned in to a two-room "modestly furnished" apartment at 328 Chauncey Street in Brooklyn for a visit with Ralph and Alice Kramden and their upstairs neighbors, Ed and Trixie Norton. The set itself—derived from Gleason's own childhood Brooklyn flat—was virtually television *noir*, a bare-bones room stocked only with icebox, stove, sink, window, battered dresser, and a wooden table covered with a checkered tablecloth.

The characters who inhabited this bleak house provided all of the color necessary. Gleason's Ralph Kramden dominated the show. He was the loser who never learned from his mistakes, and who in fact eagerly remade them week after week, a Don Quixote of perpetual failure, tilting at windmills with get-rich-quick schemes for diet pizza or Day-Glo wallpaper. Usually he railed against life to wife Alice, played by Audrey Meadows:

RALPH: *Alice, I'm gonna belt you one.*
ALICE: *Oh, you are, are you? Well, go ahead and belt me, Ralph.*
RALPH: *One of these days, Alice. One of these days—POW—right to the moon!!*

The only reason Alice could have married and stayed with such a loudmouth lout had to be love. Ralph had some redeeming characteristics: for one thing he was so sincerely obnoxious that he became almost endearing. There was also a compelling pathos in Gleason's creation of Kramden as a Depression-bred busdriver who aspired to be a millionaire: the man who would be king never left his stark little rooms in Chauncey Street.

Television, as one historian has pointed out, was Gleason's redemptive *deus ex machina*. Until television his career had more or less plodded along. He had made a few desultory B pictures for Warner's in the early 1940s and then returned to New York. His big break came when he worked in the musical comedy *Follow the Girls*, which ran on Broadway for 882 performances. One scene, in which the 250-pound Gleason was disguised as a WAVE, provoked enough notice that he landed a job on radio. Two years later he was tapped to play Riley for the new television series "Life of Riley." Then in 1951 he began starring in "Cavalcade of Stars," where he developed such characters as the Poor Soul, Joe the Bartender, and Reggie Van Gleason, 3d. It was here that he created the "Honeymooners" sketch, the single most important impetus for elevating him beyond being just another brassy video comic.

Gleason's own show began on CBS in 1952, and his rise was meteoric. *TV Guide* voted him the best comedian of the year. The show opened with the June Taylor dancers doing a Busby Berkeley production number; "The Great One" (an appellation given him by Orson Welles) would then come on stage, escorted by the Glea Girls. Sipping a cup of "tea," he would mug for the camera for awhile, call for "a little travelin' music," and then execute a storklike rolling sidestep to exit on the line "And awa-a-a-y we go!!"

From October 1955 to September 1956, "The Honeymooners"—TV's first spin-off—was a half-hour show on its own, with Gleason as Ralph, Audrey Meadows as Alice, Art Carney as Ed Norton, and Joyce Randolph as Trixie Norton. Filmed before a live audience at the Park-

Gleason as Joe the Bartender.

CBS Inc.

"The Honeymooners": Ralph Kramden (Jackie Gleason), Ed Norton (Art Carney), Alice Kramden (Audrey Meadows), and Trixie Norton (Joyce Randolph).

Wisconsin Center for Film and Theater Research

Sheraton Hotel in New York, the show had a distinctly theatrical tone, with the audience applauding first entrances of principal characters and performers doing an inordinate amount of audience-milking (count the double-takes).

Ralph always seemed to drag Ed Norton into the thick of his schemings, like a lamb to the slaughter. But Norton never got the axe, and he had a dignity that helped to exaggerate Ralph's bumptiousness.

Some of the best scenes took place between big dreamer Ralph and iconoclast Alice. After she had popped his latest balloon, Ralph would rage, "Oh, you're a regular riot, Alice! Har-de-*har*-de-har-har!" But by the end of the program, he would embrace her and say, "Baby, you're the greatest!"

Alice could give it back, too. In one episode Ralph came in, eyes aglow with yet another scheme: "This is the biggest thing I ever got into." And Alice retorts, "The biggest thing you ever got into was your pants."

How sweet it was.

Ernie Kovacs 1919–1962

He was on television for little more than a decade, and he flitted from network to network even at that, but in his bursts of comedic genius Ernie Kovacs left an impact that still makes the small screen vibrate. He was one of the few performers in TV's Golden Age who had not arrived full-blown from vaudeville or Broadway. The master of pure television comedy, Kovacs pioneered in tapping the medium's technological potential. His lavish use of electronic trickery created a humor possible only on television.

Kovacs specialized in unswerving lunacy. Variously characterized as "zany," "a buffoon," and "bizarre," as well as "a madcap Nihilist" and "a Dadaist," he admitted that "No one else is mad enough to go in for all this nonsense." He once had a set built entirely on a fifteen-degree angle, and then he had the camera adjusted fifteen degrees so that when he poured himself a cup of coffee, the liquid seemed to defy gravity as it shot off at a crazy (fifteen-degree) angle. In a variant of this, Kovacs had a set built upside down: he then had the camera inverted, so that he appeared to be walking across the ceiling. When he poured water from a pitcher, it poured upward.

With Kovacs television became a forum for the suspension of reality: visual images cavorted to music, whether they were celery sticks breaking or eggs plummeting into a skillet or perhaps gentleman germs discussing the Cold War. All were choreographed to favorite Kovacs music, such as that by Deems Taylor or Béla Bartók; on a lucky night there would be an appearance by the Nairobi Trio or a reading by the Martini-swigging poet, Percy Dovetonsils. Once Kovacs whacked a golf ball solidly into the camera's eye. Following a shattering crash, the screen went black and not a sound was heard until Kovacs purred, "And let that teach all of you out there to pay attention."

Kovacs debuted on television in Philadelphia, where he had been hired to do a cooking show. An extant clip shows Kovacs slapping around a head of lettuce

Ernie Kovacs caught in mid-skit.

NBC

for being "smart" with him. Gradually he developed his characters, and such gags as the wristwatch that boomed out with Big Ben chimes. In 1951 he did "It's Time for Ernie," followed by "Ernie in Kovacsland" and others. "The Ernie Kovacs Show" originated over NBC from New York, but Kovacs never found a permanent network slot and spent much of his time making guest appearances on other people's programs. In 1957 *TIME* magazine hailed him as "that rare being, a home grown product of TV"; *LIFE* called him an inspired personality who took "advantage of the enormous flexibility of the TV camera" and put together situations "so outrageous yet so skillfully done that his nonsense takes on a life of its own."

Kovacs's meanderings into television's inner workings set him apart. His most famous sketch was a half-hour pantomime in which a character named Eugene wandered through a world inhabited by a sculpture garden of embracing, panting stone lovers; when he took a copy of *Camille* from the bookshelf, it emitted a consumptive cough; the *Mona Lisa* gave him a leering, come-hither chuckle.

Kovacs's wife and costar, Edie Adams, called his humor the "back of the head" variety—unfettered and visual. Kovacs taught us not to take the screen literally. Though he exploited the medium's potential as no one else had, he never found a niche. Maybe the wonder is that he ever got turned loose on the small screen at all. What could the networks expect from a performer whose favorite fan letter read, "Dear Sir: You idiot"?

Ernie Kovacs as Percy Dovetonsils.

CBS Inc.

Steve Allen born 1921

Multitalented Steve Allen was the original host of the "Tonight Show" in the mid-1950s. His concept of a program rooted in a comedy format, but with much ad-libbed patter, music, and celebrity guests, evolved from his radio and television days in the 1940s, when he was everything from a host for a daily radio comedy show to a late-night disc jockey, to a tongue-in-cheek commentator for wrestling matches.

In June 1953 Allen began the "Tonight Show" as a local night-owl show on the NBC flagship station in New York, WNBT; in September 1954 the program went national on NBC, airing live nightly from 11:30 P.M. to 1:00 A.M. Allen's "Tonight Show" was truly an exercise in experimental TV. The program usually began with Allen seated at the piano, improvising and chatting amiably. He would then move over to his desk to banter with regulars Gene Rayburn and Skitch Henderson or with guest stars; new discoveries Eydie Gorme and Steve Lawrence might sing, or Allen might embark on one of his ventures-with-mike into the studio audience. He had devised this while working in radio: when a guest failed to show one night,

Steve Allen was the original host of NBC's "Tonight Show."

NBC

leaving Allen stranded with thirty minutes of airtime to kill, he grabbed a standing microphone and headed for the audience. It worked so well that he continued to use the audience for informal ad-lib comedy, and it became a popular part of the "Tonight Show." Other regular features included the Question Man (who gave the punch line before the straight line), Stump the Band, and remote broadcasts from the "man in the street" outside the NBC studios.

Allen presided over the "Tonight Show" until January 1957. By then he was already six months into his prime-time variety-comedy hour, "The Steve Allen Show," which was pitted not only against the "Ed Sullivan Show" but against "Maverick." The ratings battle among these programs in 1956–1957 was historic, and "The Steve Allen Show" held its own. One of the most gifted comedy troupes on television inhabited this show: Don Knotts, whose answer to the famous question, "Are you nervous?" always provoked a hysterical "NO!!"; Bill Dana as Jose Jiminez; Pat Harrington as the Italian golfer Guido Panzino; and Louis Nye as the suave Gordon Hathaway.

The protean Allen was composer of the "Tonight Show" theme, as well as "This Could Be the Start of Something Big" and four thousand other songs and lyrics. In his informal, likable manner and in the breadth of his interests and talents, Steve Allen pioneered late-night television with a stylishness yet to be surpassed.

Jack Paar born 1918

In July 1957 Jack Paar took over the "Tonight Show" from Steve Allen, establishing his own brand of late-night entertainment based on informality, sentimentality, and unpredictability. Listeners warmed to his desk-and-sofa "talking heads" format, not because it was an infinitely interesting arrangement in itself, but because Paar was never boring. Thin-skinned? You bet. Slightly "blue"? Sometimes—it kept Priscilla Goodbody, the NBC censor, on her toes. Droll? Excessive? Of course.

Paar—like Dave Garroway, Steve Allen, Arthur Godfrey, and Garry Moore—was one of a new breed of professional television-age entertainers bred in broadcasting. Paar left school at age sixteen and got a job as a radio announcer. After serving in World War II, he played minor roles in some Hollywood films and then became a summer replacement for such radio stars as Jack Benny and Arthur Godfrey. By 1952 he had his own TV show—"Up to Paar"—on NBC. This program, as well as a CBS "Morning Show," fizzled. But with the "Tonight Show" Paar found his audience. Surrounded by a band, singers, and guest stars, Paar held forth from a set made up of an old desk and an L-shaped sofa. Regulars included Dody Goodman (until she got too popular), Elsa Maxwell, Cliff Arquette, Genevieve, and announcer Hugh Downs.

"The show is nothing," Paar once said. "Just me and people talking. Historic naturalness. We don't act, we just defend ourselves." Paar had what he himself called "my cute little Presbyterian face," and could make the most ephemeral subject matter seem important if he felt like it; if he didn't he could be devastating. And temperamental: he once quit on the air, and tears flowed as frequently as did such trademark phrases as "I kid you not!"

One of Paar's great secrets was to let people see human beings as human, including himself. Mr. Heart-on-the-sleeve didn't put on a show—he was the show. As critic Tom Shales has pointed out, Paar's days on the "Tonight Show" represented days "when to get on a talk show, you had to have something to say."

Paar on the "Tonight Show" with Cliff Arquette and Genevieve.

Wisconsin Center for Film and Theater Research

Jack Paar, who became host of the "Tonight Show" in 1957.

India ink on artist board by Al Hirschfeld. The Margo Feiden Galleries, New York City

The Rise of Broadcast Journalism

Live coverage of news-making events started in broadcasting's earliest days. The first commercial news broadcast began at 8:00 P.M. on November 2, 1920, when KDKA in Pittsburgh reported the outcome of the presidential election in which Warren G. Harding defeated James M. Cox; in 1921 the heavyweight championship bout between Jack Dempsey and Georges Carpentier was transmitted from ringside, and the World Series was broadcast. At first, such instant access to information was as much a novelty as early radio itself. Radio news was sporadic at best, lasting a minute or two except for special events and usually consisting of announcers reading headlines from the newspapers.

The print media had been among broadcasting's earliest supporters. But in the early 1930s, as advertisers turned increasingly from newspapers to radio, collegiality degenerated into a press-radio war. In 1933 wire services effectively banned all network use of their facilities. The result was that NBC and CBS established their own news-gathering organizations, and their evolution as news media began.

With the onset of war in Europe in the late 1930s, the networks vastly expanded their news bureaus, and radio became the most vital communications medium in the world. One measure of radio's status was that when President Franklin Roosevelt set up the Office of War Information in June 1942, CBS news commentator Elmer Davis was chosen to head this domestic and international clearinghouse and coordinating agency for wartime news and information. But this Golden Age of radio news mined itself out in the years after the war, and between 1948 and 1960 television became the preeminent stage for broadcast journalism.

Television's visual dimension greatly enlarged the boundaries of our lives. It became increasingly difficult to consider one's identity in strictly local terms, as television began to document—and to a large extent came to provoke—the emergence of what Marshall McLuhan termed "the global village."

Edward R. Murrow in London.

CBS Inc.

A 4201
OXFORD
CIRCUS

Graham McNamee 1888–1942

In the earliest days of radio, Graham McNamee carved out a historic niche for broadcast announcers. His opening, "Good evening, ladies and gentlemen of the radio audience," was one of the best-known trademarks from the early 1920s until his premature death in 1942.

McNamee started out at WEAF in New York in 1923. Hired as an announcer because of his clear, vibrant speaking voice, he attracted wide attention for his broadcasts of the 1923 World Series, where he not only reported on the game but on the whole scene: spectators, weather, ballplayers themselves. "The average fan wants excitement," he said. "The secret of imparting enthusiasm lies in conveying grandstand or ringside atmosphere to those listening at home." By 1925 his reporting of the World Series was so popular that he received more than fifty thousand fan letters—though some complained that he deviated too much from the game: "What do we care whether you are hot or cold, wet or dry; or what's the state of your health?"

McNamee was one of the first announcers to use his own name on the air; others, even Norman Brokenshire, who would himself became a popular announcer, were forced to use call letters (Brokenshire, for example, was known as "AON"). But McNamee became a legendary figure. He was the first to broadcast American political conventions, notably the 1924 Republican convention in Cleveland, where he conveyed the excitement of the Coolidge bandwagon rolling to victory. By 1927 McNamee was the best-known announcer in the industry—a position capped by his coverage that year of Charles Lindbergh's triumphant return from Paris.

McNamee's informal, chatty style made him one of radio's big stars in the 1920s. But with the advent of commercials and the consequent need for closer timing and a less free-wheeling style, McNamee was pushed aside by others in sports announcing. He did remain as radio's most popular general announcer through the 1930s, when he brought his gift of gab and his delight at being a "straight man" to announcing such programs as Rudy Vallee's "Fleischmann Hour" and Ed Wynn's "Texaco Fire Chief" show.

McNamee's contagious enthusiasm and ability to perform and inform under nearly any circumstances helped pioneer the field of broadcast journalism.

Graham McNamee interviewing Babe Ruth at Yankee Stadium.

NBC

Graham McNamee broadcasting during the fire aboard the Normandie, *February 9, 1942.*

Broadcast Pioneers Library

Red Barber born 1908

Radio had a natural affinity for sports. One of the finest sports announcers ever to broadcast was Walter Lanier "Red" Barber, "The Verce of the Brooklyn Dodgers"—who was actually a blond, self-styled "country boy" from the South.

Dean of radio's sportscasters Red Barber is shown here early in his career, broadcasting in his wedding suit.

Photograph courtesy Beverly Frick

Barber began broadcasting major-league baseball in 1934 with the Cincinnati Reds. Five years later he left for Brooklyn and stayed there. His soft drawl instantly identified Barber to millions, though his dispassionate style was quite a contrast to the subjective bent of other sportscasters. "The Old Redhead," as he called himself, put disinterested accuracy above all other Barber principles. He thought of himself as a reporter, without bias, and never even admitted to being a Dodgers fan. Brooklyn fans, though first confused by the drawl, were soon won over by his colorful, non-hysterical reporting. He would choreograph his play-by-play with nuance rather than bombast, pausing slightly just before a fly ball was caught, or raising his voice a bit and increasing the tempo for a close play.

Barber was clinically thorough in preparing for games, having memorized—in the antediluvian pre-computer days—the Spalding *Guide Book* back to 1899. He also kept charts on every major-league player and would use these details to flesh out the game, giving it life and personality.

Barber was several times voted best baseball announcer in the country and won an adulatory following that heaped up to five hundred letters a week on him. His favorite was a letter from a man who wrote: "My wife was a semi-invalid. She enjoyed your broadcasts. Yesterday she began to sink, but she heard the last out. She died happy."

H. V. Kaltenborn 1878–1965

A pioneer radio news commentator, H. V. Kaltenborn was affectionately called "the suave voice of doom" for his sonorous, authoritative—not to say affected—style of broadcasting. Kaltenborn's roots were in print journalism, where he had increasingly specialized in foreign news coverage. He also built up a reputation as a lecturer on current events, a facility that led to his broadcast debut on April 4, 1922, in which he discussed a coal strike for his New York listeners—thereby becoming "radio's first news analyst." One of Kaltenborn's great talents was in extemporaneous speaking, and his broadcasts were usually based on a few quickly jotted notes.

Kaltenborn continued to develop his knowledge of foreign affairs in the 1920s as a correspondent variously based in Russia, the Far East, and amidst the Spanish Civil War. During that war Kaltenborn was described as "a solid, substantial figure in a well-cut business suit and pince-nez, Phi Beta Kappa key dangling across his ample vest front; but as a slight concession to the adventurous nature of his trip he wore a steel helmet."

Kaltenborn's finest hour—and certainly a turning point in radio journalism—was the Munich crisis of 1938. On September 12, from Studio Nine at CBS in New York, Kaltenborn broadcast Hitler's invasion of Czechoslovakia, and then kept vigil until the end of September, when Neville Chamberlain returned to England from Munich and announced "peace for our time." For eighteen days Kaltenborn's clipped voice described conditions in "Yirrup" in nearly one hundred broadcasts. In between broadcasting he slept on a cot; with each new bulletin he would go on the air instantly with a commentary. NBC news chief A. A. Schechter said, "You could wake Kaltenborn up at 4 o'clock in the morning and just say, 'Czechoslovakia'—one word—and he'd talk for thirty minutes on Czechoslovakia." One wag called him Columbia's "gem of the ozone."

Kaltenborn's performance radically altered the status of the radio news commentator and helped give radio a credibility as an information source that it had previously lacked. For the first time, one critic said, "history has been made in the hearing of its pawns." Americans found that the documentary quality of being present at history-in-the-making was of overwhelming fascination: radio became the prime source for foreign news.

H. V. Kaltenborn, the "suave voice of doom."

Pastel on paper by Winold Reiss. H.V. Kaltenborn Collection, The Mass Communications History Center, State Historical Society of Wisconsin

Lowell Thomas 1892–1981

Lowell Thomas's broadcasting career began in 1930 out of pure happenstance. He had already established himself as a popular travel lecturer when William Paley hired him to broadcast CBS's evening news program. Thomas won an instant following and, switching to NBC three years later, became an American institution, introducing each of his programs with the familiar "Good evening, everybody," and closing with "So long until tomorrow."

Thomas's news commentaries were read in a style almost neighborly. Given the hour of his broadcasts—6:45 to 7:00 P.M.—he tried to make his reports lively but palatable. "I am on the air when people are getting ready for dinner or just having dinner, or just finishing dinner. I never felt it was my responsibility to destroy the digestive system of the American people." Instead he presented a folksy news digest. His first broadcast, on September 29, 1930, was prototypic:

"A procession of German Fascists was attacked today by Communists in the town of Unterbermsgruen. . . . Adolf Hitler, the German Fascist leader, is snorting fire. . . . A cardinal policy of his now-powerful German party is the conquest of Russia. That's a tall assignment, Adolf. You just ask Napoleon."

Thomas was never one to denigrate his contribution to broadcasting, once exclaiming that "the voice of Lowell Thomas probably has been heard by more people than any other voice in history—including those of FDR, Winston Churchill, Hitler, and Mussolini." Yet he also knew that he was not, strictly speaking, a journalist. What he did was to report news events in his own way. He always thought of himself as an entertainer first, "just as Bob Hope and Bing Crosby are entertainers." It was a magic formula, and Thomas's forty-six-year tenure as a working newscaster was the longest in network radio history.

Lowell Thomas broadcasting at Gangtok, Sikkim.

CBS Inc.

In 1930 CBS founder William S. Paley hired Lowell Thomas to broadcast the network's evening news.

CBS Inc.

Walter Winchell 1897–1972

"Good evening, Mr. and Mrs. North America and all the ships at sea!" With this familiar bark Walter Winchell regularly opened his Sunday-night radio broadcast. For the next fifteen minutes Winchell would bounce up and down in his old armchair, his soft felt hat pushed back on his head, his tie loosened, and yell out "Flash!" into the microphone while frenetically jiggling a telegraph key. At the end, he would slather off his Jergens-sponsored program by sending to all and sundry "lotions of love."

Winchell invented the modern gossip column when he was hired by the *New York Evening Graphic* in 1924; his first column was called "The Newsense," setting the pattern for his unabashed fabrication of such portmanteau words as "oomph." This "greatest town gossip in the greatest town in the world" had begun as a vaudevillian, performing with Eddie Cantor, George Jessel, and others in the "Newsboys Sextet" in 1910. He then became a "keyhole reporter," contributing tidbits to the *Vaudeville News*. In 1929 he took his gossip column to the *New York Mirror*, which became a springboard to Hearst syndication and radio.

In the mid-1930s, at the height of his popularity, Winchell's column was read by an estimated seven million readers, and his radio broadcasts—slightly milder versions of his column—reached an estimated twenty million listeners. Grist-for-the-mill tidbits were sent by phone, mail, wire, or to Winchell's table at the Stork Club, where he held court. His car was equipped with a siren, shortwave radio, and red lights so that he could always be "on the spot."

The "pillar of gab" was a contentious and controversial commentator who invented a slangy gossipese in part to hinder litigious victims of his column. Though guilty of conducting vendettas and provoking fierce rivalries, Winchell was the unquestioned king of gossip in its Golden Age.

Walter Winchell, radio's king of gossip.

India ink on artist board by Al Hirschfeld, circa 1955. National Portrait Gallery, Smithsonian Institution

Edward R. Murrow 1908–1965

Edward R. Murrow guided the formative years of radio and television journalism, as one pundit said, like "a defrocked Bishop," using the microphone as his bully pulpit. With his documentary eye for detail and sepulchral voice, he set the style for broadcast journalism—and virtually invented the genre of news reporter known as "broadcast journalist."

Murrow joined CBS's fledgling news division in 1935, before radio news reporting had charted its course. As Europe moved toward war, he was sent to London to hire a small staff and prepare programs for short-wave broadcast. He had already hired William R. Shirer when German tanks rolled into Austria in March 1938. CBS decided to let Murrow and Shirer broadcast their reports of the crisis themselves, and William Paley quickly orchestrated a dramatic round-robin broadcast of CBS reporters spread across Europe; on March 13, Robert Trout in New York introduced the CBS "European News Round-Up," with reporters broadcasting live from London (Shirer), Paris (Ed Mowrer), Rome (Frank Gervasi), Berlin (Pierre Huss), and Vienna (Murrow). In the first of more than five thousand broadcasts he would make during his career, Murrow reported, "It's now nearly 2:30 in the morning and Herr Hitler has not yet arrived. No one seems to know just when he will get here, but most people expect him sometime after 10:00 o'clock tomorrow morning." The Round-Up—which won wide acclaim and was frequently repeated as events warranted—was not only a turning point for Edward R. Murrow, marking his initiation as an on-air journalist, but proved a milestone in the evolution of broadcast journalism itself. In the months to come, Murrow assembled the journalistic corps that would set the standard for broadcast news thereafter, including Eric Sevareid, Charles Collingwood, Winston Burdette, and Howard K. Smith.

Murrow's broadcast style reflected his sense of the reporter as an instant documentarian of atmosphere and events. When he broadcast from London in the

Edward R. Murrow

Oil on canvas by Guy Rowe, 1957. Edward R. Murrow Collection, Fletcher School of Law and Diplomacy, Tufts University

fall of 1939 and early 1940 during the blitz, he became—in the words of Sevareid—"a kind of Boswell to a great city's trial by fire." Murrow's London broadcasts were both impressionistic and graphic: "In many buildings tonight people are sleeping on mattresses on the floor," he reported after weeks of air raids. *I've seen dozens of them looking like dolls thrown by a tired child. In three or four hours they must get up and go to work just as though they had a full night's rest, free from the rumble of guns and the wonder that comes when they wake and listen in the dead hours of the night.*

World War II quickly established itself as a radio war, bringing Americans instant access to events thousands of miles away. From 1938 until the end of the war, radio was the most vital communications medium in the world. It was Edward R. Murrow's somber "This . . . is London" that best symbolized radio's transformation into a vast engine of information. Archibald MacLeish told him that his broadcasts "destroyed the suggestion that what is really done beyond 3000 miles of water is really not done at all." Americans could no longer assume themselves beyond the pale.

Ed Murrow Slaying the Dragon of McCarthy

Brush drawing on paper by Ben Shahn, circa 1955. William S. Paley

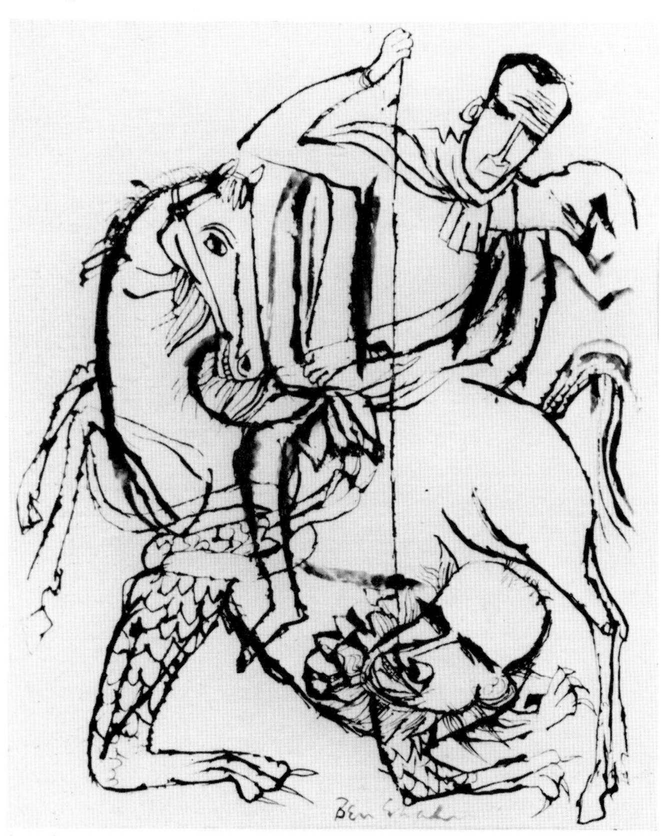

In the years following the war, as television news began to take shape, Murrow was again in the forefront. With Fred W. Friendly—with whom he had co-produced "Hear It Now" for radio—he launched "See It Now," a pioneering effort that became the prototype for television news documentaries. The first prime-time broadcast of the program (April 20, 1952) began with simultaneous shots of the Atlantic and Pacific oceans; Murrow and Friendly said, "We thought that a medium capable of doing this was capable of providing reporters with an entirely new weapon in journalism." "See It Now" was a forum for insight into difficult issues, ranging from the threat of the atomic bomb to the first reports linking cigarette smoking to lung cancer (with Murrow puffing away as he reported), to the case of Milo Radulovich, an Air Force officer who had been stripped of his rank because of his family's political activities. Don Hewitt, now executive producer of "60 Minutes," has said that television was never the same after the first broadcast of "See It Now." Until then, leading radio journalists had scorned television as a lowbrow medium: "But from that moment on, they began to see television in a different light—as a place that had room for their kind of broadcasting as well as for an Ed Sullivan or a Lucille Ball."

The most riveting of the "See It Now" programs was that of March 9, 1954, devoted to Senator Joseph McCarthy—"told mainly in his own words and pictures." McCarthy, then still at his demagogic height, was filleted by those words and pictures and soon toppled from power.

McCarthy became the victim of the medium he had so successfully exploited, thereby (and inadvertently) proving Murrow's basic premise about the neutrality of the camera's eye: "It will broadcast filth or inspiration with equal facility and will speak the truth as loudly as the falsehood. It is, in sum, no more or less than the men who use it."

Fred W. Friendly (left) with Edward R. Murrow, preparing for a "See It Now" broadcast.

John Daly born 1914

At 5:47 P.M. on April 12, 1945, the International News Service teletype in CBS's Studio 28 rang out a ten-bell news flash: "WASHN—F.D.R. DEAD." John Daly, the CBS news correspondent and analyst who was in the studio preparing for his regular 6:15 broadcast, saw the bulletin and immediately signaled for the technician to cut in on the network program in progress ("Wilderness Road") and to give him a live microphone. By 5:49 P.M. Daly was on the air across the country, breaking the news that "A press association has just announced that President Roosevelt is dead. All that has been received is that bare announcement. There are no further details as yet, but CBS World News will return to the air in just a few minutes with more information as it is received in our New York headquarters."

This was the first announcement that Americans heard of Roosevelt's death, and as more copy flashed across the wires, Daly—without script or notes—narrated the unfolding story, smoothly and spontaneously describing one of the major news events of our time.

His ability to ad-lib masterfully was one of the things that had won Daly his CBS White House billet in October 1937, only months after the twenty-three-year-old first ventured into radio news on WJSV, CBS's Washington affiliate. His most important prior employment had been with the Capital Transit Company, where he was a schedule engineer—or, as he later described it, a person who counted buses. People kept asking him if he was a radio announcer, so Daly decided that he might as well find out. CBS hired him, and he soon found himself covering the White House. His first assignment was a Fireside Chat, for which his colleagues helpfully put him through an etiquette drill: wear a white shirt, sedate suit and tie, bow slightly and say, "How do you do, Mr. President," and then slink quietly into the background. Daly comported, and even bothered to rehearse slinking backward. President Roosevelt greeted him with, "Welcome to the gang. Remember, I'll be your severest critic," and then proceeded to borrow the young reporter's watch.

After five years of covering the White House, Daly joined the CBS staff in London in 1942. During the war he reported on the North African and Italian campaigns, was the first to describe the bombing of Monte Cassino and to give an eyewitness account of the fall of Messina, and covered as well the surrender of the Italian fleet at Malta. He returned to the United States for the 1944 political conventions, winning a great scoop by disclosing FDR's decision to choose Senator Harry S Truman as his new running mate over the incumbent Vice-President, Henry Wallace. After the war Daly returned to Europe for such stories as the Nuremberg trials and the Berlin airlift. Back home he gave on-the-spot reports on everything from presidential campaigns to the 1947 Texas City disaster and the 1949 Florida hurricane. "After all my years of reporting, I can't miss," he once said. "For years, if there was a penguin arriving at the zoo, I was the one who'd stand there with a mike waiting for the egg to hatch"—while wearing white tie and tails "for the sake of harmony."

Beginning in 1947 Daly also hosted "CBS Is There," a dramatic program presenting such historic events as the Salem witch trials but using the guise of modern, on-the-spot news coverage. In 1953 he was appointed vice-president of television news, special events, and public affairs at ABC. While he continued to be master of ceremonies on the highly popular CBS game show "What's My Line?" he also anchored ABC's nightly newscasts. Although Daly at first had been wary of television, convinced that the camera could tell the listeners far more than any amount of his ad-libbing, he relented and tried the medium after a friend pointedly asked, "How do you know you'll look silly on TV? You've never been on it." In fact the television camera captured Daly's insouciance wonderfully, and he became a familiar figure on the small screen.

For his direction and participation in the coverage of the Korean War, the 1952 and 1956 political conventions, the Army-

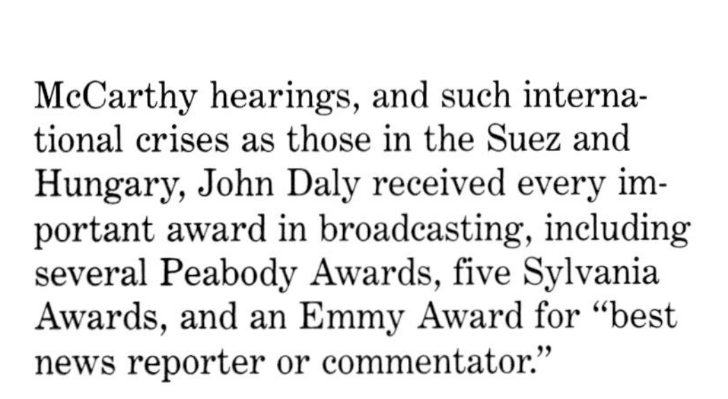

Emmy Award presented to John Daly in 1954 for "best news reporter or commentator."

Virginia Warren Daly

McCarthy hearings, and such international crises as those in the Suez and Hungary, John Daly received every important award in broadcasting, including several Peabody Awards, five Sylvania Awards, and an Emmy Award for "best news reporter or commentator."

John Daly, drawn by James Montgomery Flagg as Daly was reporting Franklin Roosevelt's death from the CBS newsroom in New York.

Charcoal on paper, 1945.
Virginia Warren Daly

John Cameron Swayze
born 1906

John Cameron Swayze pioneered television's nightly news, serving as the so-called commentator when NBC-TV premiered its fifteen-minute evening newscast—"The Camel News Caravan"—in 1948. A former actor and newspaperman, Swayze combined a friendly, animated manner with a brisk delivery, punctuating the program's tempo with such trademark phrases as, "Now let's go hopscotching the world for headlines."

Swayze would begin each newscast with a crisp recital of the latest bulletins, while the camera featured footage from Washington, London, or Rome. After more news film, and possibly recorded interviews with political headliners, the show would switch to Chicago for the weather forecast. Sports scores were scribbled on a chalkboard.

Swayze's imperturbability and quite remarkable memory held it all together. He enjoyed concentrating on human-interest stories, and when he ended his program with a cheery, "That's the story, folks. Glad we could get together!" his nightly audience of five million probably nodded in agreement. Swayze once said, "Leaving people feel good—that's my role."

John Cameron Swayze hopscotched the world for headlines on NBC's "Camel News Caravan."

NBC

Walter Cronkite born 1916

Walter Cronkite was the man for whom the term "anchorman" was coined. The occasion was not when he became managing editor of the "CBS Evening News" in 1962, but a decade earlier, when he was chosen to "anchor" the televised 1952 national political conventions on CBS.

Long before he earned such titles as "Uncle Walter" and "Old Ironpants," Cronkite had a solid journalistic career that spanned newspaper, wire service, radio, and television reporting. He worked on the *Houston Press* in 1935–1936, moved to Kansas City as news and sports editor and broadcaster for radio station KCMO, and then joined United Press. One of the first reporters accredited to accompany American forces after Pearl Harbor, he gave eyewitness accounts during the war of such events as the first United States bombing raid over Germany, the combat landing in North Africa, and the invasion of Normandy; he was dropped with the 101st airborne into the Netherlands and covered the Third Army in the Battle of the Bulge and the German surrender of northwest Europe. From 1946 to 1948 he was United Press's chief correspondent for Russia before returning to Washington, where he delivered the six o'clock evening news—on WTOP—ad lib. "In the beginning," he said, "I wanted to end every broadcast saying, 'For further details, see your local newspaper.' " Cronkite joined CBS in 1950, where he became known as "a reporter's reporter." His 1952 anchoring of the political conventions set a new standard for TV journalism. He also became well known for his smooth-as-silk hosting of the CBS docudrama series, "You Are There," in which he "reported" carefully re-created historical events.

Walter Cronkite

Tempera on board by Robert Vickrey, 1966. National Portrait Gallery, Smithsonian Institution; gift of Time, Inc.

Cronkite has spoken out vociferously against the trivialization of broadcast journalism by blow-dried types who spent their college years, as he has put it, taking "Trench Coat I and II and Makeup I and II." He came up through a tradition of "gut journalism" and still believes that television news ought to be done informally. "The audience says, 'Come in and tell me the news today.' " It's the way he learned as a young reporter, when his boss at United Press would call for a report every night at midnight. "He wanted to know what was going on. Bam. Bam. Bam. Five minutes of news. It was the best training in the world. You had to be ready for that call every night."

Chet Huntley 1911–1974
David Brinkley born 1920

In the late 1950s American television news was dominated by the NBC team of Chet Huntley and David Brinkley. The two had first been paired to cover the 1956 national political conventions and had achieved such acclaim that the network kept them yoked for the NBC evening news program.

"The Huntley-Brinkley Report" soon became the top-rated news show, although in format it did not differ substantially from the CBS News program with Douglas Edwards: each was a meager fifteen minutes long and focused on the anchor reading headlines, covering a few stories in more detail, and using some film footage. Neither NBC nor CBS had begun to set up more than skeletal independent news-gathering organizations; local affiliates had to be depended upon for footage and stories. This type of leisurely newsreel approach continued until the early 1960s, when CBS doubled the number of its domestic network bureaus and gave itself the capability of covering almost any event in the country on its own. In 1963 CBS—quickly followed by NBC—expanded its nightly newscast to thirty minutes, and television news suddenly acquired a depth it had never before achieved, except in its coverage of such special events as political conventions or inaugurations.

Chet Huntley, the laconic West Coast newsman, was a good counter to the puckish and more acerbic David Brinkley. But it was Brinkley who invented a new broadcast style: rather than the "voice of doom" school, he represented what one critic called a "pungent and economical style of prose . . . and a dry, sardonic tone of voice which carries great authority." His knack for the succinct phrase, which served him so well during convention coverage while other correspondents felt it necessary to babble incessantly, pointed out one of his greatest strengths—he knew when not to talk. On the other hand, he could talk extemporaneously, he once estimated, for "perhaps four or five hours" on the air.

"The Huntley-Brinkley Report," while still trapped by network prejudice for entertainment over news in the 1950s, helped to turn that particular tide by its very popularity. The program's coda—"Good night, Chet," "Good night, David"—became a part of the national lexicon.

Chet Huntley and David Brinkley preparing for election-night coverage.

Broadcast Pioneers Library

Dave Garroway 1913–1982

From "Garroway at Large" in 1949, through a decade that saw the inception and perpetuation of the "Today" show and three years of "Wide, Wide World," Dave Garroway brought his fine hand—really, his upraised palm—to the craft of broadcasting. This maestro of the medium was, as critic Tom Shales has written, born to television.

"Today" host Dave Garroway.

NBC

"Garroway at Large" was broadcast from Chicago, which in the late 1940s was a renaissance city for creative television programming; Fred Allen said, "They ought to tear down Radio City and rebuild it in Chicago and call it Television Town." Garroway's half-hour show was low-keyed live entertainment with a standard for wit and literacy that marked the "Chicago school" of broadcasting.

When NBC's Sylvester "Pat" Weaver cast about for a host-communicator for his new morning program, the news-entertainment "Today" show, Garroway auditioned "in my best March-of-Time doom-heavy voice, and it got me in. Come to think of it, I never have used that voice on the show." Weaver's orders for Garroway were to "wake America, wash it, dress it, give it breakfast and send it to work."

It was a radical thought in 1952—television at 7:00 A.M.—but Garroway's courtly personality was perfect. "Today" colleague Jack Lescoulie said, "Dave was the ideal man to talk to people early in the morning. He moved like molten glass—slow and easy." Although it took a while to catch on, the "Today" show soon won a loyal following. Most Americans were up by 7:00 A.M., and they learned to flip on their TV sets for Garroway. Lescoulie, Frank Blair, Betsy Palmer, and a prima-donna chimpanzee named J. Fred Muggs—who bit everyone, including the host—rounded out the cast, but it was Garroway the interlocutor who set the show's tone.

On the first broadcast, he said, *Well, here we are. And good morning to you—the very first good morning of what I hope and suspect will be a great many good mornings between you and I. Here it is . . . January 14, 1952, when NBC begins a new program called "Today" and—if it doesn't sound too revolutionary—I really believe begins a new kind of television.* Three hours later he signed off with his hand raised in the peace gesture.

Relentlessly curious, seemingly never without the right word or phrase, he seemed at ease with the medium's intimacy with its audience: "The lens seemed to be so direct and friendly, really," he once said, "almost as if I could see somebody there. It was a black channel to the people."

He was also unflappable—when he was doing a (live) commercial demonstrating a shockproof watch, he dropped it and it shattered. So Garroway just shrugged, and with total aplomb said, "Well, that didn't work, did it?"

"Wide, Wide World" was another milestone Garroway program on NBC. A Sunday afternoon show, its purpose was to illustrate how television could obliterate distance. Cameras were strapped to roller coasters, or Garroway and crew would navigate "American Waters" from the Mississippi to the Grand Canyon to New York harbor. At the end of the program he would quote Edna St. Vincent Millay:

The world stands out on either side,
No wider than the heart is wide;
Above the world is stretched the sky,
No higher than the soul is high.

It was pure Garroway. As always.

Garroway on the first broadcast of the "Today" show.

NBC

Variety

More than any other kind of programming, variety reflected the quintessentially mass character of commercial broadcasting. In the showmanship tradition of P. T. Barnum and Flo Ziegfeld, variety shows on radio and television offered a panoply of programs ranging from music to amateur hours to game shows. Variety's formula of mixing different kinds of performance had deep roots in American theater history, extending back to days when minstrels, burlesque, and opera were often offered on the same playbills.

The idea of presenting variety in entertainment clearly proved successful from broadcasting's earliest years. By 1934 variety programs ranked first in audience popularity and scored among the top three shows until 1940. During World War II, top variety performers were the most in demand to entertain America's overseas troops, appearing with great success in war bond drives, USO tours, and on the Armed Forces Radio Service (AFRS) in such programs as "Command Performance." Game shows and amateur hours won wide followings at home. With these shows, broadcasting became a national participatory sport. After the war four of radio's top five programs had a variety format. By the early 1950s televised variety had won such large audiences that programs like Milton Berle's were credited with making commercial television a success. Ultimately, the key to variety's popularity was its sheer American facility to provide a bit of something for everyone.

CBS Radio Theatre marquee advertising "Major Bowes' Original Amateur Hour."

CBS Inc.

CBS
RADIO THEATRE

CARLOS TAP
KOBELEFF BALLET
CARLOS TAP

CHRYSLER CORPORATION
PRESENTS
MAJOR BOWES ORIGINAL
AMATEUR HOUR

CBS RADIO THEATRE

HOTEL BRYANT
CAFE-BAR
RESTAURANT

MEN'S SHOP

Clicquot Club Eskimos

The Clicquot Club Eskimos began broadcasting in 1925 from WEAF in New York. Led by Harry Reser, they affected Eskimo-style fur garb—even before they had a studio audience—and played a mélange of banjos, brass, and bells. Reser composed their theme, "The Clicquot March," which was used to introduce the program—the first original theme to be identified with a radio show. The announcer would open the show with "Look out for the falling snow, for it's all mixed up with a lot of ginger, sparkle, and pep, barking dogs and jingling bells and there we have a crew of smiling Eskimos." And the band would play its theme. The announcer would then continue: "After their long breath-taking trip down from the North Pole, the Eskimos stop in front of a filling station for a little liquid refreshment—and what else would it be, but Clicquot Club Ginger Ale—the ginger ale that's aged six months. Klee-ko is spelled C-L-I-C-Q-U-O-T."

Sponsored by Clicquot Club ginger ale, the Clicquot Club Eskimos dressed in costume even before they had a studio audience.

From *Great Radio Personalities in Historic Photographs* by Anthony Slide, Vestal Press, 1988

In 1929 the band moved to a six-hundred-seat Broadway theater previously occupied by the Ziegfeld Follies. The audience was greeted with a stage set depicting an aurora borealis and a large mechanical husky that jumped from behind a cardboard iceberg and barked along with the opening theme. During the show the Clicquot Club Eskimos' music ran strongly to such songs as "Barney Google," "Yes, We Have No Bananas," and "Ain't She Sweet."

At first they installed an enormous six-ton soundproof "glass curtain" between the band and the studio audience, to keep audience noise from cluttering the broadcast. But the five rows of hot lights quickly baked the Eskimos, so the curtain was raised, and they soon discovered that the radio audience enjoyed hearing the live audience applaud, laugh, and cough.

The Clicquot Club Eskimos enjoyed the celebrity status that radio was beginning to give to its favorite stars; "personalities" had begun to dominate the air. This popularity rubbed off on the Clicquot Club's product, proving, as a 1929 NBC booklet exclaimed, that "even a ginger ale may be personalized and dramatized," and that "Broadcast Advertising can change a brand-name, difficult to pronounce and spell, into a household word." Though the Clicquot Club had at first hesitated to enter the radio market—there were, as NBC admitted, few "guide-posts along this new highway of advertising" in 1925—they had decided to take the chance. By 1929 the company was convinced that "this new advertising medium is one of unlimited possibilities." Broadcast advertising had in fact become big business.

Rudy Vallee, 1929.

Wisconsin Center for Film and Theater Research

Rudy Vallee 1901–1986

Known variously as "the troubadour of the ether" and "the matinee idol of flapperdom," Hubert "Rudy" Vallee hosted "The Fleischmann Hour," one of radio's most popular programs from 1928 through the 1930s. As an adolescent he had taught himself to play the saxophone by listening to the records of Rudy Wiedoeft—whose first name he adopted—and later worked his way through the University of Maine and Yale by freelancing in dance bands. After playing with a band at London's Savoy Hotel, he took "My Time Is Your Time" as his theme song, formed his own orchestra—the Connecticut Yankees—and opened at the Heigh Ho Club in Manhattan. During his stint in 1928 at the club, Vallee developed both his trademark greeting ("Heigh Ho, everybody!") and his characteristic use of the megaphone to sing above the sound of the orchestra.

His radio career began at the Heigh Ho Club that year, broadcast over WABC, and he proved such a sensation that in the fall of 1929 he was signed to do "The Fleischmann Hour" on NBC.

The success of this weekly variety program—usually credited as being the first variety show on radio—produced a cascade of show-business entertainment on the air. The Depression and "talkies" had forced vaudeville off the stage and into radio studios, and Vallee's "Fleischmann Hour" replaced the Palace Theater as vaudeville's primary booking agent. Vallee himself described his show's purpose: "to discover and develop more personalities and stars than any other radio show before or since."

The program would open with a band number, followed by a comedian's routine, a novelty act, a dramatic spot, another novelty act, and a "leave 'em laughing" finale. At least once during every show, Vallee would sing his most-requested song, "I'm Just a Vagabond Lover":

For I'm just a vagabond lover,
In search of a sweetheart it seems,
And I know that someday I'll discover her,
The girl of my vagabond dreams.

Fans, often teenaged girls, would swoon and shriek at his live stage performances when he sang "The Maine Stein Song" or "Cheerful Little Earful." One of the points that all this clamor suggested was the astounding ability of radio to popularize personalities. Vallee himself had never played the Palace before starring on "The Fleischmann Hour," and when he did, following his first broadcasts, he was virtually besieged. Clearly, something powerful had transpired between audience and performer on the air.

Perhaps Vallee's greatest contribution during his ten-year tenure on "The Fleischmann Hour" was to help introduce such entertainers as Kate Smith, Burns and Allen, Bob Hope, Eddie Cantor, Joe "Wanna Buy A Duck?" Penner, and, most preposterously, the ventriloquist Edgar Bergen. The program was consistently ranked first or near-first among all radio programs during the 1930s, a result probably more of Vallee's likability than his crooning. "I never had much of a voice, and it was all in my nose," he once confessed. "But I think one reason for the success was that I was the first articulate singer—people could understand the words I sang. And at least I had pitch."

Radio's first national matinee idol, Rudy Vallee.

India ink and wash on paper by Miguel Covarrubias, 1930. Prints and Photographs Division, Library of Congress

One of Vallee's most popular songs was "I'm Just a Vagabond Lover."

Alice Becker Levin

Eddie Cantor 1892–1964

Banjo-eyed vaudevillian Eddie Cantor got his start on Broadway in the Ziegfeld Follies from 1917 to 1919. Billed as the "Apostle of Pep" in the unsurpassed 1919 Follies, Cantor worked almost entirely in blackface, a routine he had first adopted in 1912 when he "blacked up" with burnt cork for a Coney Island vaudeville revue. During the 1920s he had such big Broadway hits as *Kid Boots* (1923) and *Whoopee* (1928).

Backed by his stage fame, Cantor made his radio debut on Rudy Vallee's "Fleischmann Hour" in February 1931. The following September he launched his own program, "The Chase and Sanborn Hour," featuring Jimmy Wallington as announcer-straight man, Burt Gordon as the Mad Russian, Harry Einstein as Parkyakarkas, the violinist Rubinoff, and singers Bobby Breen and Deanna Durbin. Dinah Shore and Burns and Allen would also make their radio debuts on Cantor's show.

One of Cantor's innovations was to use a live studio audience. Listeners responded by making "The Chase and Sanborn Hour" radio's top-rated program, bumping off "Amos 'n' Andy" in 1933. For the next two years, it was estimated that more than 50 percent of the radio audience tuned in to this show. One critic noted that Cantor's appeal—typified by his theme song, "I'd Love to Spend This Hour with You"—was that he made listeners feel as if they were "no longer alone in the empty living room; they were part of the show."

Cantor's rise to success from a humble Lower East Side upbringing, where he sang on street corners and in saloons, impressed the Depression audience, and his blend of optimism and patriotism struck a receptive chord. But Cantor also capitalized on radio's verbal appeal and gauged his performances beyond the studio. It was one of the reasons for his sustaining power—why he lasted, while entertainers like Ed Wynn, who stuck mainly to highly visual old vaudeville routines, faded from listener grace after the mid-1930s.

Eddie Cantor broadcasting in the early 1920s.

National Museum of American History, Smithsonian Institution

Finding it impossible to work in a broadcast booth with only a microphone standing inertly in front of him, Cantor pioneered the use of live studio audiences. Cantor is seen here during his first Camel broadcast, March 1938.

CBS Inc.

Cantor was also part of television's infancy in the early 1950s, bringing what one critic called his "pure song salesmanship and personality" to the "Colgate Comedy Hour." He said that he had waited until 1950 to make his television debut because in that year "5,000,000 more television sets were sold . . . after all, Cantor likes to play to full houses."

Jessica Dragonette
circa 1905–1980

In the heyday of radio's infatuation with operetta and musicals, Jessica Dragonette reigned as the "Sweetheart of the Air." She began her broadcasting career in the 1926–1927 season on radio's first singing-acting serial, as "The Coca-Cola Girl." She sang on "The Philco Hour"—NBC's pioneering program of light opera—from 1927 to 1930 and then reached full stardom in her Cities Service concert series from 1930 to 1937.

Radio's penchant for light classical music and musical comedy was inspired by Hollywood's craze for musicals during the Depression, and Jessica Dragonette's sweet, clear voice established her as broadcasting's counterpart to Jeanette MacDonald. She herself was dazzled by this "new concept in entertainment," as she wrote in her autobiography, *Faith is a Song*: "By the amplification of the voice radio carried to the largest imaginable audience the most intimate form of expression." Critics raved about her as "the girl with the dimple in her voice," and wondered at radio's strange power over the imagination: "How the unseen singer, product of the machine," said the *Philadelphia Inquirer*, "could blossom into favor as human and personal as that of any star who won popularity . . . is the interesting phenomenon unfolded in this story."

Dragonette's "Cities Service Hour" was fairly serious stuff for a mass audience: of the ninety-seven shows broadcast, seven were operatic, six were adaptations of film musicals, and all the rest were operettas or Broadway musicals. A major source of its popularity—at its

Jessica Dragonette with Rosario Bourdon's orchestra on "The Cities Service" program. Announcer Ford Bond is in the center.

Broadcast Pioneers Library

Jessica Dragonette was radio's golden lady of song in the 1930s.

Photograph by Ray Lee Jackson from *Great Radio Personalities in Historic Photographs* by Anthony Slide, Vestal Press, 1988

peak the program reached sixty-six million people—was the lavish musical talent presented, including a forty-piece orchestra, a twenty-voice chorus, and such well-known singers from the Metropolitan Opera as Gladys Swarthout, Rose Bampton, and James Melton.

From 1935 to 1942 Dragonette was voted the "Queen of Radio" by the readers of *Radio Guide*. Bob Hope once said, "Jessica probably has done more for radio than anyone else."

Major Edward Bowes 1874–1946

By the mid-1930s radio had begun to mine out its golden supply of talent, and frantic searches were undertaken to discover "new" Rudy Vallees, Kate Smiths, and Eddie Cantors. Into this slough of despond strode Major Edward Bowes, a kindly voiced, fatherly impresario of the air who inaugurated his "Original Amateur Hour" in 1935. He promised to "open the flood-gates of the air" and it worked: amateurs flocked to auditions by the thousands, and the radio audience made amateurism a blockbuster craze. The year it premiered, "Major Bowes' Original Amateur Hour" was fed to more than sixty stations of the NBC Red network and was voted the most popular show on the air.

While the contestants performed, Major Bowes sat enthroned behind a microphone. He would spin the wheel of fortune: "around and around she goes, and where she stops, nobody knows." The audience telephoned in its vote, and winners were dispatched on a kind of post-amateur theater circuit. To the losers went the gong—the equivalent of the old vaudeville hook—which the Major banged with a wooden mallet. One contestant was a miner who had devised a glockenspiel-like headdress that he played by wiggling his ears. Another favorite was a young Manhattan Dead Ender who wanted to raise money for a tombstone for his father's grave. He had looked several over in the graveyard, but "dey all had names on 'em already." The most promising winners over the years included a young Beverly Sills, and a singing group called the Hoboken Four—among whom was Frank Sinatra.

In 1939 Major Bowes was described by *TIME* magazine as heavy-jowled but stylish: "He still has most of his finely spun orange-blond hair, and skin that seems to have been massaged, steamed and lotioned for days. From any angle his nose is mightier than Jimmy Durante's."

Major Bowes—his title derived from a World War I reserve commission—came upon a spontaneous new use for radio, sparking an amateurism vogue that consistently kept his program in the top ten and has kept wheels spinning to this day.

Major Edward Bowes, July 1936.

CBS Inc.

Minnie Pearl born 1912

After studying theater at Nashville's Ward-Belmont College, Sarah Ophelia Colley began working in musical and minstrel shows throughout the South. One day, while staging a play in Alabama, she met a woman who became the prototype for "Minnie Pearl," the good-hearted small-town spinster who was to become a renowned figure in American humor. Dressing in gingham and a flat-brimmed hat that prominently dangled its price tag, "Minnie Pearl" would come out on stage with an enormous "How-DEE!," sing burlesque versions of country songs in a wonderfully cracked voice, and tell anecdotes about family life in fictional Grinder's Switch.

In 1940 Minnie Pearl joined the cast of the "Grand Ole Opry," and her career flowered when she began appearing on the "Prince Albert Show," the NBC network portion of the Opry. The Opry itself had begun in 1925 with broadcasts from Nashville's WSM Studio B; the announcer, George Hay, gave the program its name. Most performers in the earliest years were singers or "pickers" (of guitars, banjos, mandolins, fiddles, and zithers), with someone occasionally playing a "wind" instrument—a harmonica. Among the groups were the Fruit Jar Drinkers and the Gully Jumpers; the first real singer was "the Dixie Dewdrop," Uncle Dave Macon. In the 1940s the Opry grew to more than one hundred performers, including Roy Acuff, Ernest Tubb, and Hank Williams.

Minnie Pearl toured with Acuff, PeeWee King, and Eddy Arnold during the war, and in 1946 began a twelve-year radio collaboration with Rod "Boob" Brasfield. Rather than rely on visual humor and slapstick, Pearl and Brasfield tailored their comic routines to radio, emphasizing verbal humor and quick repartee. Long a favorite on the Opry—one of radio's most successful programs—Minnie Pearl was elected to the Country Music Hall of Fame in 1975.

Minnie Pearl

The Country Music Hall of Fame and Museum

Minnie Pearl in full swing with Roy Acuff (far right).

The Country Music Hall of Fame and Museum

Bob Hope born 1903

In a 1948 article *Coronet* magazine said of Bob Hope: "For a onetime song-and-dance man who was frequently compelled to accept second billing to Siamese twins and trained seals, Leslie Townes (Bob) Hope has managed to do pretty well for himself." Born in England and raised in Cleveland, Hope got into show business after high school, when a call was put out for several acts to fill a bill headed by Fatty Arbuckle at a Cleveland theater. He and a partner, George Byrne, worked up a dance routine as "Those Dancemedians," which they subsequently toured around Midwestern vaudeville houses. A typical exchange in the act went: Q: "Why is a pig's tail like getting up at 5 o'clock in the morning?" A: "Twirly, Mr. Byrne, twirly." Hope then formed his own vaudeville company, which included Edgar Bergen and Charlie McCarthy, before leaving for a successful Broadway career that began with the 1932 musical *Ballyhoo* and was followed by *Roberta* in 1933.

Bob Hope

Oil on canvas by Norman Rockwell. Bob Hope

Though he had once refused to appear on radio because it "would never amount to anything," Hope made his first radio appearance on Rudy Vallee's "Fleischmann Hour" in 1933. His crackling, one-liner patter suited radio, and he soon became one of the most popular performers on the air. His radio success led to an appearance in the film *The Big Broadcast of 1938*—a largely forgettable movie worth remembering only because it was here that Hope introduced what became his theme song, "Thanks for the Memory." His biggest radio success came later that year, with "The Pepsodent Show." For the next fifteen years, Hope would glide into America's living rooms on "The Pepsodent Show," bringing along sidekicks like Jerry Colonna, orchestra leader Skinnay Ennis, and the vocal group Six Hits and a Miss.

When World War II began, Hope tried to enlist but was told that he could better serve as an entertainer. With his USO troupe he made trips to Sicily, Britain, Alaska, North Africa, and the South Pacific. He was the mainstay of the Armed Forces Radio Service (AFRS), the network of stations that the army and navy set up in the European and Pacific war theaters, and he performed regularly on such AFRS shows as "Command Performance," "Mail Call," and "GI Journal"—programs heard only by service personnel.

In Alaska for a grim Christmas visit, he told the troops, "Be happy, you guys. Be proud. You know what you are—you're God's frozen people!" Another time, while stopping off in the Philippines, he was approached by a sergeant who asked him to entertain his outfit, too. Hope at first said that it couldn't be squeezed in but agreed when the sergeant urged, "But it's *got* to be now." So, at nine o'clock the next morning, Hope and troupe landed on a tiny Marine runway and had the soldiers roaring with laughter for more than an hour. The next day, most of the Marines died on Guadalcanal.

The indefatigable Hope made his television debut in 1950 on the "Star Spangled Revue," and he and his specials have been woven into the American fabric

Hope on stage at Munda airstrip, Munda, New Guinea, August 5, 1944.

National Archives

Hope in London to entertain troops with Frances Langford and Jack Pepper, June 28, 1943.

National Archives

Bob Hope on USO tour in the 1940s.

Bob Hope

ever since. For years, Christmas meant another Hope show with the troops—he was always there. Writer John Steinbeck once said, "It is impossible to see how he can do so much, can cover so much ground, can work so hard and can be so effective." Millions of people over several generations have a lot of memories to thank him for.

Bing Crosby 1904–1977

On the third day out of New York aboard the S.S. *Europa* in June 1931, a relaxing William Paley overheard a recording of "I Surrender, Dear" wafting from a phonograph in a nearby stateroom. The unfamiliar baritone voice intrigued him, and he immediately sent a radiogram back to CBS: "Sign up singer named Bing Crosby."

Harry Lillis Crosby started out as one of the Rhythm Boys (along with Al Rinker and Harry Barris) for the Paul Whiteman band in the late 1920s. After being discovered by Paley, Crosby debuted on CBS in September 1931; convinced that he had flopped, he wired his brother, "I gave all I had, and it's no good." But CBS was deluged with wires to the contrary, and Crosby soon found himself a national sensation.

His laid-back, unassuming singing style was called "crooning" by everyone but Crosby himself, who considered that too syrupy a characterization. He just called himself a baritone. Radio and recordings combined to make him a household voice, and his list of early hits reads like a litany of the best of American popular song: "Stardust," "Dancing in the Dark," "Wrap Your Troubles in Dreams," "I Found a Million Dollar Baby in a Five and Ten Cent Store," "How Deep is the Ocean," "Temptation," "Home on the Range," and his theme song, "Where the Blue of the Night Meets the Gold of the Day."

Crosby's best-known radio program was the "Kraft Music Hall," which began in 1936 and remained one of radio's top-rated shows for ten years. His casual, conversational style proved vastly likable, and his supporting cast included the Jimmy Dorsey band (later John Scott Trotter took over), Mary Martin, comic Bob Burns, and guests from opera (Feodor Chaliapin, Lauritz Melchior), the stage and concert halls (Duke Ellington, Art Tatum), and Hollywood. It was an extraordinarily eclectic mix, but it worked. Crosby biographer Barry Ulanov said, "The longhairs were shortened; the crew-cuts were lengthened. Men and women from the opera and concert stage were humanized; jazz musi-

Bing Crosby

CBS Inc.

Bing Crosby singing for the troops at the Stage Door Canteen in London, August 31, 1944.

National Archives

cians and rowdy comics were treated with dignity. . . . The result was a humanization all around."

During the war Crosby was one of the leading performers on bond tours and overseas shortwave broadcasts for the Armed Forces Radio Service (AFRS), and at service camps and hospitals. After an AFRS broadcast from London, in which Crosby spoke German for the Germans, a journalist cabled home how "Der Bingle" had emerged as a first-rate American propagandist. A group of servicemen formed a Hope and Crosby campaign for President and Vice-President; he was presented with the "G.I. Oscar" based on a poll taken by *Yank*, the official army newspaper, and was chosen as "the person who had done more for the morale of overseas servicemen than any other entertainer."

After the war Crosby created a precedent in broadcasting by winning the right to transcribe his show—that is, to prerecord his radio show on tape rather than to broadcast it live. His argument was that it would give the program a fluidity previously lacking—mistakes and miscues could be edited out—and guests could appear where schedule conflicts before would have made it impossible. The *New York Times* called the 1946 premiere of the prerecorded Crosby show "portentous": "Mr. Crosby has delivered a major if not fatal blow to the outworn and unrealistic prejudice against the recorded program." The shows were put on sixteen-inch discs (three to a half-hour program) and sent to radio stations around the country. For the first time, a show scheduled for 8:00 P.M. in New York could also be broadcast at 8:00 P.M. in Los Angeles. The only problem was that the discs were being pressed too quickly, with resulting glitches in tonal quality. A backroom employee of Crosby Enterprises, John Mullin, remembered that he had discovered a new German recording system during the war; with the German surrender he dismantled two Radio Frankfurt machines and mailed the parts and fifty rolls of plastic tape to himself at his San Francisco address. By October 1947 ABC was in business broadcasting the Crosby show on tape.

Crosby premiered on television singing in "The Red Cross Program" in February 1951. Though critics said, "Bing's relaxed style and easy-going ways were made to order for home viewing," he himself favored doing occasional specials rather than a regular show. By then, radio—and a neatly dovetailing recording career that brought sales of more than 300 million records—had established Bing Crosby as the most famous popular singer of his time.

Kate Smith 1909–1986

She was once called "radio's own Statue of Liberty." Franklin Roosevelt introduced her to King George VI and Queen Elizabeth by saying, "This is Kate Smith—this is America."

Kate Smith began her show-business career in Broadway musicals cast as the fat-girl foil for comics like Bert Lahr. When Ted Collins became her manager in 1930, though, all of that was changed, and she concentrated on singing. Her radio debut came on May 1, 1931, in a fif-

Kate Smith

India ink and ballpoint pen on artist board by Herman Perlman, circa 1935. National Portrait Gallery, Smithsonian Institution

teen-minute program over CBS. For this program she introduced the song that would become her theme—"When the Moon Comes over the Mountain." For the rest of the 1930s, she was a ubiquitous radio presence, displacing Rudy Vallee as broadcast's number-one attraction and becoming known as "the Queen of Radio."

In 1938 she won exclusive rights from Irving Berlin to sing his national paean, "God Bless America"—a song that he had written twenty years before but had put away as being too "sticky." Miss Smith's performance of this song week after week made it the unofficial national anthem, particularly after the outbreak of World War II.

During the war Kate Smith tirelessly performed for the troops in the United States and Canada and did a phenomenal job of raising money for war bonds. She traveled nearly 520,000 miles and raised $600 million in bonds—more than any other single person. In one marathon radio bond drive, she collected $105,392,700, the largest sum ever amassed in a single effort.

She continued her radio work following the war and made her television premiere in 1950. At one point in the mid-1950s, she had an hour-long TV program in the afternoon, an hour variety show on Wednesday nights, and a forty-five-minute daily radio program at noon.

She never tried to be the perky, girl-next-door kind of singer; she just went on stage and sang out in her vibrant contralto. There was no one else like Kate Smith. From the 1930s to the 1950s, she was America's songstress.

Kate Smith with World War II naval troops.

CBS Inc.

Horace Heidt 1901–1985

Every Tuesday night on NBC, from 1939 to 1941, Horace Heidt spun a great numbered wheel on "Pot o' Gold" to select the telephone listing of someone, somewhere in the United States. He then—live, on the air—called that number, awarding up to a thousand dollars to those who answered his call and his questions. They didn't even have to be listening to "Pot o' Gold" to win—they simply had to answer their phones; needless to say, one thousand dollars was a substantial amount in these closing years of the Depression! Despite the eighty-million-to-one odds, so many people stayed home on Tuesday nights, poised between radio and phone, that restaurants and theaters reported pronounced drop-offs in business. Some theater owners offered one-thousand-dollar prizes to anyone called while they were at a movie.

Heidt, already known as the orchestral leader of "Horace Heidt and His Musical Knights," would whirl the wheel three times during each show. The first number spun selected the volume to be taken from "Pot o' Gold's" enormous collection of telephone directories; the second spin gave the page number to be turned to; the third pinpointed the lucky—not to say unsuspecting—contestant. Between spins and whirls, Heidt and his Musical Knights would play dance numbers until announcer Ben Grauer shouted, "Stop, Horace!" and informed the voice out there somewhere, on the other end of the line, that a thousand dollars would be coming his way via Western Union.

"Pot o' Gold" had the dubious honor of being the first program to give away money to listeners across the country. Given its sensational reception, the show clearly knew its audience.

Bandleader and game-show host Horace Heidt.

Photograph by Ray Lee Jackson. Broadcast Pioneers Library

Groucho Marx 1890–1977

Leader of a zany band of brothers who first called themselves "The Four Nightingales," Groucho Marx wisecracked his way through vaudeville, radio, movies, and television for forty years. He and his brothers got their nicknames from a monologuist named Art Fisher: Groucho for his saturnine temperament, Harpo for his musical instrument, Chico for his reputation with women, Gummo—who left the act in 1918—for his gumshoes, and Zeppo for Zippo, the star of a chimpanzee act.

Groucho began wearing his swallowtail coat in a sketch called "Fun in Hi Skule," where he played the professor. The ubiquitous cigar came about (a) because he liked cigars, and (b) because they were useful as a prop if you missed a line—you just blew smoke until you thought of something to say. The moustache, his chief trademark for thirty years, was at first only greasepaint.

The Marx Brothers scored a great success on Broadway in the 1924 production *I'll Say She Is*, thanks largely to drama critic Alexander Woollcott's ecstatic review. In addition to stage hits of *The Cocoanuts* and *Animal Crackers*, the boys appeared in the subsequent film versions, as well as in such other classic movies as *Horsefeathers*, *Duck Soup*, and *A Night at the Opera*.

Groucho's career as a single began with the "You Bet Your Life" quiz show on ABC radio in 1947; in 1950 the show moved to NBC and television. The quiz format allowed Groucho to do what he did best—ad-lib outrageously. Aided by announcer George Fenneman and, of course, The Duck, Groucho would engage the unwitting contestant in banter: when he asked one woman how old she was, and she said, "approaching 40," the gallant Groucho replied, "From which direction?" Or, interviewing a tree surgeon, he asked, "Have you ever fallen out of one of your patients?" Or, faced with a contestant frozen with mike-fright: "Either this man is dead, or my watch is stopped."

Groucho reached the height of his popularity with his 1950s TV show, though the only real difference between "You Bet Your Life" on radio and television was that television allowed the vision of Groucho puffing away on his cigar and The Duck dropping down peremptorily. Here was the essential Groucho, stripped down and uncluttered. The viewer could almost see Groucho's self-styled epitaph flash subliminally across the screen: "I hope they buried me near a straight man."

Groucho Marx posed with his wheel of fortune and The Duck in 1957, when "You Bet Your Life"—then in its ninth season—was the longest-running TV show on the air.

Broadcast Pioneers Library

Groucho Marx

Lithograph by Al Hirschfeld. The Margo Feiden Galleries, New York City

Mark Goodson

Gouache on paper by Don Bevan. Mark Goodson

"What's My Line?" cast, headed by John Daly and featuring panelists (from the top) Arlene Francis, Bennett Cerf, and Dorothy Kilgallen.

From the original caricature by Al Hirschfeld. Virginia Warren Daly

Mark Goodson born 1915

Mark Goodson was the pioneering impresario of television game shows. His greatest contribution was the panel show "What's My Line?" which began in 1950. The moderator was John Daly, who had a distinguished career as a journalist and anchorman. The four panelists, after an initial shakedown period, were columnist Dorothy Kilgallen, actress Arlene Francis, publisher Bennett Cerf, and a fourth panelist slot that alternated such guests as David Niven and Steve Allen.

Goodson began his career in radio. He teamed up with Bill Todman in the mid-1940s and originated such popular radio programs as "Winner Take All," "Stop the Music," and "Hit the Jackpot."

"What's My Line?" which debuted in February 1950 and soon became a Sunday-night staple, was as much a celebrity get-together as a game show. The biggest names in show business and politics appeared as the "Mystery Guest," including Nöel Coward, Judy Garland, Paul Newman, Bette Davis, Frank Sinatra, Bob Hope, and Chief Justice Earl Warren. The witty and urbane John Daly would ask the "Mystery Guest" to "enter and sign in please" on a chalkboard, while the panel sat blindfolded. The questions and answers that followed—with guests trying to mystify the panel with disguised voices and dissembled answers—produced some of television's most engaging moments.

Although "What's My Line?" quickly spawned such other Goodman gold mines as "I've Got a Secret" and "Two for the Money," it was this initial venture that became one of television's most popular and longest-running programs. Daly's determination to make "What's My Line?" a serious game show, along with its celebrity roster, elevated the show to a national institution. Broadcast live fifty-two weeks a year, it came on at 10:30 Sunday nights and, as producer Gil Fates has written, it "changed a nation's habits. The accepted way to finish the weekend was with *What's My Line?*, *The CBS Television News*, and bed."

Arthur Godfrey 1903–1983

In the decade from 1941 to 1951 Arthur Godfrey was a ubiquitous broadcasting presence, filling local and network airwaves with homespun palaver from early morning until evening prime time. In 1947 *Newsweek* labeled him "CBS's most valuable personality property—and the freshest voice on the generally stale 1947 [radio] network air."

Godfrey, who had appeared on a 1929 talent show as "Red Godfrey, the Warbling Banjoist," began his radio career at WFBR in Baltimore and then became a staff announcer for NBC in Washington, D.C. Fired for sarcastically ad-libbing commercials—a trait that would become one of his trademarks—he joined CBS's Washington station, where his program was picked up in New York one night by Walter Winchell. Winchell liked him and gave his show a plug in his column: *Godfrey is stuck down there across the Potomac from the Capitol. But he is big-time. His quips are sly—and his fly-talk is terrifically Broadway or Big Town. Some shrewd radio showman should bag him for New York to make our midnight programs breezier. . . . I haven't picked a flop yet.*

Bolstered by this publicity, Godfrey began an enormously successful freelance career and for the next several years appeared on such radio shows as the Chesterfield program, "Professor Quiz," the "Arthur Godfrey Show," and an early morning program on WCBS. In 1948 his "Talent Scouts" began to be televised. People seemed to respond to Godfrey as if he were the good old boy-next-door; for his folksiness he was once called "the Huck Finn of radio" by Fred Allen. In his peak years he was cuddled by the press as "the Old Redhead" and kidded for his near-deification: "It is only a matter of time," exclaimed one trade magazine, "until the second syllable of 'Godfrey' will be forgotten."

But Godfrey's sentimental reputation was blasted away on October 19, 1953, when he fired singer Julius LaRosa on the air. The engaging LaRosa had committed a cardinal sin against Godfrey by arranging his own business deals, hiring

Arthur Godfrey

Gouache by Al Hirschfeld. The Margo Feiden Galleries, New York City

an agent, and signing an independent recording contract. So, on the October 19 program, Godfrey noted that LaRosa was on the verge of becoming "a great big name" and asked him to sing "I'll Take Manhattan." Then, as the show was ending, Godfrey said to the audience, "Thanks ever so much, Julie. That was Julie's swan song with us; he goes now out on his own, as his own star." Godfrey later said that LaRosa had violated the unwritten rule of the show, that "we have no stars . . . we're all one family," and that LaRosa "lacked humility"—words that would boomerang squarely back to Godfrey himself, who came across in the episode as an overweening egoist. The television eye had stripped away Godfrey's "Old Redhead" veneer, and though he remained on TV for several more years, he was never able to recapture the public confidence, much less its affection.

Arthur Godfrey with Julius LaRosa.

CBS Inc.

Milton Berle born 1908

Strange things began to happen in America beginning in September 1948. On Tuesday nights at 8:00 P.M., a pall would descend: theater sales would slack off, restaurants would thin out, and neighborhood movies would play to empty houses. And where had Americans mysteriously gone? Home to their new television sets, one and all. And what magnificent magnet drew them to their small screens? Why, it was "Uncle Miltie," bounding into their living rooms by dog sled or parachute, juggling or doing card tricks, and perpetrating outrageous imitations of Superman, Santa Claus, and Carmen Miranda.

Milton Berle was a forty-year-old vaudevillian when his "Texaco Star Theater" made him "Mr. Television" in 1948. He had grown up in show business guided by his mother, who had whispered "Milton, be funny" to him in his crib. At age five he won a Chaplin-imitation contest; he performed in silent films with Mary Pickford and Marie Dressler, and he toured the country as a boy entertainer for the Keith-Albee vaudeville circuit in the 1920s. In 1931 he began a triumphant two-year engagement at New York's Palace Theater—vaudeville's Mecca—where he "took the theatre by storm" as Broadway's youngest master of ceremonies. Berle spent most of the 1930s and 1940s on Broadway, in Hollywood, and performing in nightclubs. He had debuted on radio in 1934 and had appeared occasionally on both CBS and NBC, but radio was not Berle's best medium. His physical slapstick style didn't transmit, and he himself said that the problem was "You have no chance to mug, or appeal to the audience with liquid brown eyes."

Then, in the summer of 1948, Berle became master of ceremonies for NBC's televised "Texaco Star Theater," and the visual Berle clicked. He became television's "first real smash hit," according to the critics. His regular "Texaco Star Theater" was launched in September, and Tuesday nights became "Berle Nights." Originally seen in twenty-four cities, either live or on two-week kinescope delay, "Texaco Star Theater" gar-

Milton Berle, "Mr. Television."

Wisconsin Center for Film and Theater Research

nered an estimated audience of 4,452,000. And, most importantly, Berle was credited with stimulating more sales of TV sets in that medium's infancy than any other single agency. People would gather at the homes of friends who had sets, or stand outside appliance stores and press their noses against the glass to watch Berle's zany antics—and then would often return to buy a set for themselves.

Called—and not always fondly—"The Thief of Badgags," Berle earned the enmity of fellow comics for his inability to resist stealing their material. Bob Hope once said, "When you see Berle, you're seeing the best things of anybody who has ever been on Broadway. . . . I want to get into television before he uses up all my material." And Fred Allen said, "He's done everybody's act. He's a parrot with skin on." The problem was that Berle always thought the material sounded better in his voice, and as a friend commented, "The guy just can't help imitating something that has entertained."

Berle insisted on having control of every aspect of his show, from gags to costumes, choreography to camera shots, script writing to musical arrangements. He often led the band over the conductor's head during rehearsals and, as one television critic wrote in 1950, "Don't pay any attention to the gossip that four men have to thrust him into a straitjacket to keep him off the stage; I understand it only takes two men."

His combination of raucous comedy and lavish, star-studded production numbers was in direct competition on Tuesdays with Bishop Fulton J. Sheen's religious program, but as Berle said at the time, no matter: "We both work for the same boss—Sky Chief."

In 1951 NBC signed Berle to an exclusive thirty-year contract at $100,000 per year, whether his program remained on the air or not. As it happened, "The Milton Berle Show" went off the air in 1956, though he returned in 1958–1959 to star in a half-hour weekly comedy-variety show, "The Kraft Music Hall." But Berle and NBC never seemed to be able to re-create the magic of the earlier years, and "Mr. Television" thereafter rarely appeared on the small screen that he had helped to make a commercial success.

Ed Sullivan 1902–1974

When syndicated newspaper columnist Ed Sullivan premiered his variety show, "Toast of the Town," on CBS television in 1948, the critics had a field day. His smile seemed stuck on "grimace," and his awkward arm-flaying suggested some horrific torture being perpetrated by the camera's eye. One critic kindly proffered that Sullivan was "totally innocent of any of the tricks of stage presence," and Sullivan himself admitted, "I can't sing, dance, tell jokes, tumble, juggle, or train wild animals." But he did have an even rarer ability, a kind of radar for spotting talent. Once he learned to stop staring

Ed Sullivan on "Toast of the Town." Sullivan's show began in 1948 and continued on Sunday nights for twenty-three years—the longest-running variety show in television history.

CBS Inc.

Ed Sullivan

Charcoal on paper by René Bouché, circa 1955. CBS Inc.

into the camera and to focus instead on the audience, he found his métier.

Sullivan had been a successful newspaperman since the 1920s, and his syndicated *New York Daily News* column, "Little Old New York," appeared in twenty-seven other papers. He had also done some radio broadcasting, spotlighting such entertainers as Jack Benny, Irving Berlin, George M. Cohan, and Florenz Ziegfeld. "But for television," he said, "I had to learn to smile."

Sullivan's longtime connections with show business and his eye for the new and novel contributed greatly to the success of "Toast of the Town." His showcasing of established stars in the 1950s included appearances by Fred Astaire, Gertrude Lawrence, Helen Hayes, Bob Hope, and Maria Callas; but he also introduced such new performers as Dean Martin and Jerry Lewis, Margaret Truman, Elvis Presley, and the Beatles.

Retitled "The Ed Sullivan Show" in 1955, the program was a Sunday-night American habit. Variety shows had been early contributors to radio's success in its infancy, with those of Rudy Vallee, Kate Smith, and Bing Crosby each lasting for well over a decade. Television audiences in the late 1940s and 1950s responded with similar enthusiasm, and variety shows consistently topped the ratings. Sullivan's strength, as he himself said, was as a talent scout rather than as an entertainer, and he had a sixth sense about what mixture of Broadway, Hollywood, and animal acts were needed to produce a "really big shew." As Oscar Levant once said, Ed Sullivan "will last as long as other people have talent."

Nat "King" Cole

NBC

Nat "King" Cole 1919–1965

Primarily oriented to white middle-class America, programming by the major commercial networks traditionally reflected mainstream culture. Radio hosted such ethnic shows as "The Goldbergs," "Life with Luigi," and "Amos 'n' Andy" (starring two white men in burnt-cork makeup), but black participation on the networks was severely limited to the entertainment side, and then either to music or to comedies caricaturing blacks as butlers or maids.

Black musicians had played a major part in the development of popular music on radio in the 1920s. As J. Fred MacDonald has written in *Don't Touch That Dial!* early radio featured blacks in jazz groups, dance bands, religious choirs, and recorded music. By the 1930s, such performers as Fats Waller, Duke Ellington, Cab Calloway, and the Mills Brothers were radio regulars. Then, between 1935 and 1939, musical programming dropped off considerably on radio, with audiences switching their preferences to drama, news, and special events. Broadcast historian Erik Barnouw has written that, when the pendulum swung away from music in the late 1930s, blacks subsequently had a dwindling role on the air.

The economics of radio also militated against greater black participation in the major networks. Advertisers were unwilling to have their products closely identified with black performers. As Roland Marchand has suggested in *Advertising the American Dream,* commercial broadcasting helped fuel a "market democracy" whose consumer citizens were predominantly white and middle class. Radio programming consequently aimed at an audience that was defined "not in terms of the population as a whole," but in terms of the mainstream buying public.

The greatest opportunities for blacks in broadcasting came at the local level, at such stations as WJTL in Atlanta, which pioneered black news programming in the mid-1930s; at WSBC in Chicago, which was airing five-and-a-half hours of black programming a week in 1939; and at KFFA in Helena, Arkansas, which broadcast blues programming on a regular basis. Only after World War II, though, did the role of blacks on network radio change. Black participation in the war, and increasing white recognition of the implications of racism, led to a reevaluation of blacks and American society—and that reevaluation was reflected in network programming.

Nat "King" Cole's broadcasting career symbolized the partial shift in the industry's postwar attitudes toward blacks on the air. He had formed the King Cole Trio in 1937 and had made his first recordings for Capitol Records in 1943. Cole established himself as one of America's most popular recording artists with such hits as "Nature Boy," "Mona Lisa," and "Too Young." In 1946 the King Cole Trio became a successful summer replacement for Bing Crosby on the "Kraft Music Hall" radio program; Cole's appeal was ascribed to his being "a sepia Frank Sinatra." The trio was given its own radio slot that fall at 5:45 P.M. on Saturdays. Sponsored by Wildroot Cream Oil, the show remained on the air for sixty-eight weeks.

In the 1950s Cole helped break the color barrier on television. After regular guest appearances on such shows as Ed Sullivan's "Toast of the Town," Cole was given his own fifteen-minute program in the fall of 1956 on NBC, thereby becoming the first black entertainer to host his own network variety program. NBC, which could not find a national sponsor for the show, aired it without one; Cole had to pour his own salary back into production costs. By the summer of 1957 the program had increased its audience rating by 45 percent but still lacked a national sponsor. NBC kept the show on, expanding it to thirty minutes in the fall of 1957 and booking such top musical stars as the Mills Brothers, Ella Fitzgerald, and Harry Belafonte. But national sponsors still failed to materialize. Advertisers worried about the negative reaction of southern network affiliates and remained unconvinced that sponsorship of a black performer—even one with Cole's proven crossover popularity—would appeal to a mainstream audience.

After running for sixty weeks, the "Nat 'King' Cole" program went off the air in December 1957. When it came to performers, TV in the 1950s was clearly not broadcasting in black and white. As Cole himself said after his show was canceled, "Madison Avenue is afraid of the dark."

Liberace 1920–1987

Perhaps only American television could have produced the phenomenon that was Liberace. And where, except for the invention of the Las Vegas glitter palaces, would Liberace have been without television? The mass visual medium and the candelabra-ed flash were a marriage made in rhinestone heaven.

But his television days in the 1950s were salad days of relative understatement before the high-spirited "rococo vulgarity" of ermine excess—before he saluted the Bicentennial by wearing red, white, and blue hotpants and twirling a baton, and before his 1984 performances at Radio City, where he came on stage in a chauffeured Rolls Royce and danced with the Rockettes while draped in a $300,000 rhinestone-studded Norwegian blue fox cape with a sixteen-foot train. No, the television days were lower key, festooned mainly with a candelabra on a grand piano, a costume of white tie and tails, and a violin-playing brother named George.

Born Wladziu Valentino Liberace in West Allis, Wisconsin, "Lee" Liberace started piano lessons at the age of four. He learned a great deal more than "Beer Barrel Polka" as a youth, becoming proficient enough to make a solo appearance with the Chicago Symphony at sixteen. He dropped all but his last name after the manner of his idol, Paderewski, and continued to study the classical piano repertory. Then serendipity struck during a recital in La Crosse, Wisconsin, when the audience called out for him to play "Three Little Fishes" as an encore. When he did, he recalled, it "really shook 'em up," and the makings of a legend had been germinated.

He got his first national exposure in 1952, when NBC gave him a fifteen-minute weekly program. "The Liberace Show" ran for five years, was carried by more stations than "I Love Lucy," and established Liberace as television's first matinee idol. He was the biggest solo attraction in America in these years, selling out performances at Carnegie Hall and Madison Square Garden. His repertoire consisted mostly of popular

Liberace

Photograph by Herbert Georg Studio. Liberace Museum

music, old parlor songs, and classical selections—especially Chopin and Beethoven—"without the dull parts." He once explained that his audiences enjoyed a "Reader's Digest version" of classical music, and he whipped through Chopin's "Minute Waltz" in thirty-seven seconds and Tchaikovsky's First Piano Concerto in four minutes.

Forever the showman, Liberace began adopting flamboyant dress early in his career. When he played the Hollywood Bowl in 1952, he wore white tails "so they could see me in the back row." When he added a gold lamé jacket, "They crawled out of the woodwork. . . . What started as a gag became a trademark."

His ostentatious style of performing never won over the critics; one television columnist wrote that Liberace "just plays 'Lady of Spain' over and over again in different keys." But excess never led to pretense with Liberace, and after that particular critical barb, he typically responded by adding "I Don't Care" to his next performance.

Liberace never forgot that "without the show, there is no business."

Garry Moore born 1915

One of the most varied of television's variety hosts in the 1950s, Garry Moore began his broadcasting career as a staff writer for a Baltimore radio station in the mid-1930s—a job that he had landed on the recommendation of F. Scott Fitzgerald. When he was eighteen, Moore and the novelist had formed an unlikely duo to write one-act plays in the manner of the Grand Guignol cabaret melodramas popular in London in the 1920s. "Nobody bought them," Moore recalled. Later, during a stint as a news announcer and sports commentator in St. Louis, he discovered an on-air talent for comedy and changed the direction of his career.

Moore made his network radio debut in 1939 on "Club Matinee," broadcast from Chicago on NBC's Blue network. Here he worked with Ransom Sherman, legendary as the first of radio's literate comics, and Sherman's sophisticated satire helped to mold Moore's own style. NBC brought him to New York in 1942 to host a similar show there, and a year later he was hired to costar with Jimmy Durante in a CBS variety program that became one of the most popular on the air. Moore's breezy, intellectual humor proved a great foil for Durante's malapropisms, and the "Durante-Moore Show" built up a radio audience estimated at four million per week.

Moore struck out on his own in the late 1940s and developed a sixty-minute daily radio show, which CBS put on television in 1950. The "Garry Moore Show" was an hour-long early afternoon variety show that critics said set "a high standard for daytime television," and Moore soon found himself with an audience of twelve million.

Moore described his show as being run along the lines of a comic strip: "No one show of mine fractures you all by itself. But, in continuity, they go over—just like *Li'l Abner* or *Pogo*." The show resembled an old-style, Berlesque variety program: "Today," Moore said in 1953, *people think of variety as meaning songs, dancing, and jokes. In old-time vaudeville, it meant that, too, but it*

Garry Moore with Durward Kirby and Carol Burnett, June 3, 1960.
CBS Inc.

Garry Moore
CBS Inc.

meant something more. They'd have, say, a couple of hoofers, then a seal act, and then a dramatic bit with Ethel Barrymore. Why, they'd even have William Jennings Bryan speaking on free silver. That's what we're trying to do. Real variety.

Moore would open the show with a two-minute monologue, followed by a seven-minute sketch with one of the cast—such as announcer Durward Kirby—and then a specialty spot, perhaps featuring naturalist Ivan Sanderson and his menagerie of chunga birds. For the close, Moore would do a short bit of commentary or tell some carefully timed jokes. His listeners were so loyal that when he told them to send buttons to help Durward Kirby make a quilt, CBS received tens of thousands; when he sneezed on camera, the switchboard ignited with suggested remedies such as recipes for rum toddies or chicken soup.

In addition to his five-day-a-week variety show, Moore moderated "I've Got a Secret" and occasionally filled in for Arthur Godfrey on Godfrey's "Talent Scouts." But it was variety that Moore loved best. He was one of the few variety artists of radio and early television who had not come up through the ranks of vaudeville circuits, nightclub acts, or theatrical work. Moore—who called himself a "writer who learned to talk"—was a creation of broadcasting, and he succeeded because he understood that, "Above and beyond everything else, we're selling a mood, not formalized entertainment."

Drama

In the mid- and late 1930s radio found its most original voice. The dramatic works of Norman Corwin, Orson Welles, and Archibald MacLeish showcased radio as a theater of the mind, with an imaginative scope beyond that of any other arena. Made possible by network willingness to experiment with creative programming on a sustained (that is, unsponsored) basis, these dramas proved that radio—though a mass commercial enterprise—was capable of producing eloquence.

As drama had brought radio to its greatest artistic heights in the late 1930s, television's Golden Age glittered brightest in the 1950s with such live theater programs as "Playhouse 90," "Studio One," and "The Hallmark Hall of Fame." In the early years of TV, few conventions or restraints had grown up to fetter the medium's creative potential, and the finest directors, writers, and actors came to the small screen to test its possibilities. These pioneers created a vital legacy, which, though brief in years, endures as a standard.

Orson Welles

CBS Inc.

CBS

Archibald MacLeish 1892–1982

A radio play, Archibald MacLeish once explained, existed only through the spoken word. There could be no contrived stage sets or costumes or visible actors: "the word dresses the stage." With the inception of CBS's brilliant word playground, "The Columbia Workshop," in 1936, language found its mass arena. And one program stands out above all others: Archibald MacLeish's *The Fall of the City*, a verse play broadcast in 1937 that marked radio's entry onto a new plane of literacy.

MacLeish presented Fascism—the "Conqueror" in the play—as a great mailed figure, broad as a brass door, who walked toward a beleaguered town with an awful, thunderous clanging. The narrator—the "Radio Announcer," played by a then-unknown Orson Welles—tells of a crowd of ten thousand in the mythical town square, awaiting subjugation by the Conqueror:

ONE PROPHESIZED:

The city of masterless men
Will take a master.

VOICES RING OUT:

Men must be ruled!
Fools must be mastered!
Rigor and fast
Will restore us our dignity!
Chains will be liberty!

As the play ends, the crowd grovels before the Conqueror. His visor opens; the Announcer alone looks in—the brass headpiece is empty. The awfulness of this blind mass surrender leaves the Announcer helplessly murmuring:

The long labor of liberty ended!
They lie there!

Response to this play was phenomenal. William Paley later called it "an all-time pinnacle of radio drama . . . and a sensation in 1937." The "Columbia Workshop" had not only established itself as an arena for experimental radio theater but had launched a new era in broadcasting. Writers like Dorothy Parker, Sherwood Anderson, Stephen Vincent Benét, W. H. Auden, and Norman Corwin began to write for the "Workshop." *Fall of the City* had established that drama had an important place in radio.

MacLeish said that, even more than movies, books, or the stage, broadcasting came to matter over the long run because it more persistently shaped "the minds of more people than all the rest of us put together."

Archibald MacLeish

Bronze bust by Milton Hebald, 1984 cast of 1957 original. National Portrait Gallery, Smithsonian Institution

Rehearsal for The Fall of the City, *later broadcast from the Seventh Regiment Armory in New York City on March 4, 1937.*

CBS Inc.

Norman Corwin, radio's bard. Carl Van Doren once wrote that Corwin "is to American radio what Marlowe was to the Elizabethan stage."

From *Great Radio Personalities in Historic Photographs* by Anthony Slide, Vestal Press, 1988

Norman Corwin born 1910

Norman Corwin was the poet of radio's golden decade. From the late 1930s to the postwar era—while giving birth to broadcast journalism, a galaxy of popular stars from Kate Smith to Bing Crosby, and a political culture laminating a mass medium to a mass audience—radio also emerged as a showcase for words in the epic mode: words in the guise of poetry, words as drama, words in the American key. Verse and prose tapped radio's vicarious nature, the power of the imagination to embroider sound. Corwin wrote:

I am Radio
Companion of Sun and Thunder
Over the American continent. . . .
From the steaming savannahs
of the South
Up to white Alaska
From the gold-tossing cornfields
of the West
to the incredible Eastern cities
Do you hear me, America?

Corwin came to radio from journalism. CBS hired him in 1938 to experiment with a half-hour series of poetry productions. "Norman Corwin's Words Without Music" featured programs based on nursery rhymes and "word orchestrations"—or "radio poems"—based on the works of such American versifiers as Edgar Lee Masters, Carl Sandburg, and Stephen Vincent Benét. Among his best-known scripts for the program were "They Fly Through the Air"—Corwin's reaction to Mussolini's son's description of a bomb bursting into a city block as being "like a budding rose unfolding"—and the documentary "Seems Radio Is Here to Stay."

The week after Pearl Harbor was bombed, CBS, Mutual, and both NBC-Red and NBC-Blue networks broadcast Corwin's "We Hold These Truths," a strongly nationalistic play written to commemorate the 150th anniversary of the Bill of Rights, starring Jimmy Stewart and Lionel Barrymore. Corwin also produced "26 by Corwin," a six-month series for the wonderfully creative "Columbia Workshop." In 1944 CBS gave him a special series, "Columbia Presents Corwin," for which he wrote scripts ranging from documentaries to comedies. And on May 8, 1945, as war in Europe ended, Corwin's documentary ballad to the Allied victory was broadcast. "On a Note of Triumph" was quintessential Corwin, an eloquent, lyrical encomium to the triumph of freedom:

Lord God of test tube and blueprint
Who jointed molecules of dust and shook them till their name was Adam,
Who taught worms and stars how they could live together,
Appear now among the parliaments of conquerors and give instructions to their schemes.

To have heard these words broadcast must have been exhilarating, as it is to speak them aloud even today.

Norman Corwin was, as Studs Terkel has said, radio's bard: "In his giftedness, style and substance was fused."

Orson Welles 1915–1985

Orson Welles first appeared on radio in 1934–1935, in NBC's docudrama series "The March of Time," a kind of dramatized newsreel for radio sponsored by *TIME* magazine. Even before he became well known, Welles's ability to take on roles requiring any accent or age made him one of the most sought-after actors on radio; he once said that by 1935 he never earned "less than $1,000 a week as an unnamed, anonymous radio actor." In 1937 he was chosen to be Lamont Cranston, the millionaire playboy who foiled evildoers by night in the adventure serial "The Shadow." (Q: "Who knows what evil lurks in the heart of man?" A: "The Shadow knows . . . ha-ha-ha!")

Welles's theater work brought him into contact with John Houseman, and in late 1937 he and Houseman took over the tiny Comedy Theatre and ensconced their drama troupe in the newly renamed "Mercury Theatre". The company enjoyed such success that Welles persuaded CBS to hire them to present a series of plays adapted from masterpieces.

His work with the "Mercury Theatre" would be his most innovative effort on radio. Here he created—as director, writer, and actor—the quintessence of what imaginative radio drama could be. With a cast that included Agnes Moorehead, Joseph Cotton, Martin Gabel, and Welles, the "Mercury Theatre" premiered on radio with Bram Stoker's *Dracula* on July 11, 1938. Programs based on *Treasure Island, A Tale of Two Cities, The 39 Steps, Jane Eyre*, and others followed in weekly sixty-minute installments. In September the "Mercury Theatre" moved into its regular time slot, opposite the hugely popular "Chase and Sanborn Hour," starring Edgar Bergen and Charlie McCarthy; Bergen usually pulled in about 35 percent of the audience, while the "Mercury Theatre" would average 3.6 percent. And so the stage was set for one of the most bizarre events in broadcast history.

For the Halloween program on October 30, 1938, H. G. Wells's 1898 fantasy, *The War of the Worlds*, was scheduled. But chief Mercury writer Howard Koch considered the book so antiquated as to be

Orson Welles at a press conference, October 31, 1938.

CBS Inc.

laughable and set busily to work rewriting. Orson Welles joined him in the final rewrite, and somewhere along the line the key modernization occurred: use of the present tense and the addition of staccato-like news bulletins to plot the course of the Martians' progress toward Manhattan.

The god of serendipity then joined the fray, ordaining that millions of listeners would twirl their radio dials immediately following Bergen and McCarthy's opening monologue. As Bergen introduced a new and unknown singer, thwack! went dials all over America. When they tuned to the Mercury's play, many failed to realize that it was "play." They had missed Welles's warning at the beginning that it was all make-believe, and panic set in. John Houseman later suggested that the public was made especially susceptible because the Munich crisis had taken place only a month earlier. Jitters grew as the horrific creatures were described pushing their way from Grover's Mill to midtown Manhattan. Forty minutes into the program, CBS realized that all was not well, and at the break—with New York fictionally suffocating in poisonous black smoke—an announcer said, "You are listening to a CBS presentation of Orson Welles and 'The Mercury Theatre on the Air,' in an original dramatization of *The War of the Worlds*, by H. G. Wells." For the last twenty minutes, Welles narrated the denouement, with the Martians being killed by Earth's bacteria—but by then, the damage had been wrought. As broadcast historian Erik Barnouw has pointed out, the event, though a high point in radio's Golden Age of drama, was "in many ways a reenactment of *The Fall of the City*: men had rushed to prostrate themselves before an empty visor."

An unrepentant Welles later said that the hoax was possible because of radio's emergent importance: "The radio was believed in America. That was a voice from heaven, you see. And I wanted to destroy that as dramatically as possible." Though the little practical joke had exploded out of hand, all that he had actually intended was an appropriate Halloween offering: "the *Mercury Theatre*'s own radio version of dressing up in a sheet and jumping out of a bush saying boo."

Beginning in 1940 Welles's central focus was directed toward film rather than broadcasting. But in the early 1950s he became intrigued by the possibilities of television. In December 1953 he starred in Peter Brooks's TV production of *King Lear*. One critic wrote, *Like a confidently patient boxer who lets his opponent flail away for eight or nine rounds and then calmly steps in to finish the fight with one blow, Orson Welles burst onto television . . . and knocked everything for a loop. The performance he gave as King Lear established a new high for the medium in terms of power, heart and sheer ability.* Such success, alas, did not lead to a full-blown Wellesian career on television. However much he admired TV as an actor's rather than as a director's medium, he was unwilling to forego his fascination with film for the small screen.

Reginald Rose born 1920

The exigencies of television drama required the evolution of a new genre. On the one hand, live TV drama had the immediacy of both theater and radio; yet the small screen imposed constraints of its own, particularly in terms of what was seen through the camera's eye and in the unrelenting problem of time. The irony was that once these constraints had been overcome—primarily by the invention of videotape—the resulting "telefilm" productions proved less satisfactory than live drama of the early and mid-1950s.

One of the reasons that live drama worked so well was that its structure was remarkably adaptable to television. Fred Coe, producer of the "Philco-Goodyear Playhouse," explained in his 1954 "TV-Drama's Declaration of Independence" that "basically, the one-hour drama is the old one-act play, subject to the requirements of television, and endowed—in my opinion—with more scope, more depth of character and more impact." To the problem of commercial interruptions, he said that he considered commercials to be rather like curtains, which naturally broke the one-hour script into a three-act form.

One of the pioneers of live television drama was Reginald Rose, whose plays from "Studio One"—especially *Thunder on Sycamore Street* and *Twelve Angry Men*—were classically suited to the medium. Television, in fact, nurtured Rose's emergence as a major playwright: *I was being handed script-writing jobs over at CBS, on* Studio One, *while I still held on to a daytime job writing copy at an advertising agency. Eventually I had made enough of an impact to give up my day job and go into writing full-time—but in the meantime, I'd been gifted with a chance to learn my craft.*

Reginald Rose, Franklin Schaffner, and Felix Jackson during Thunder on Sycamore Street, *1954.*

Reginald Rose

Reginald Rose

Oil on canvas by Ellen Rose.
Ellen M. Rose

In one of Rose's "Studio One" plays, Art Carney starred in *The Incredible World of Horace Ford.* Carney described it as a *wonderful idea—a guy on the edge of a nervous breakdown, works in a toy factory, keeps thinking how wonderful it would be to go back to his childhood, finally, at the end, he just disappears, vanishes from the scene. Fantasy. There were all sorts of opportunities like that around, gave an actor a chance to flex his muscles.* Directors and writers were given the chance, too. Live drama flourished on television in a brief creative burst and then—like Carney's Horace Ford—disappeared.

George Schaefer born 1920

Producer-director George Schaefer helped make the "Hallmark Hall of Fame" a television legend. Beginning with *Hamlet* (starring Maurice Evans) in 1953, Schaefer produced or directed such "Hallmark" dramas as *The Corn is Green, Man and Superman, The Little Foxes, Born Yesterday, A Doll's House,* and *Magnificent Yankee*—sixty-one productions in fifteen years, with some of the greatest actors of the English-speaking stage: Maurice Evans, Judith Anderson, Eva Le Gallienne, Julie Harris, E. G. Marshall, Katharine Cornell, Mary Martin, Paul Douglas, Christopher Plummer, Art Carney, Richard Burton, and Alfred Lunt and Lynn Fontanne.

Schaefer was a master theatrical craftsman, with a strong background in producing and directing. At the New York City Center he directed and produced such plays as *The Heiress, Idiot's Delight, Richard II,* and *The Taming of the Shrew*—a repertory reflecting his belief that more classics should be presented in the popular theater. For his television debut in 1953 he was given full artistic control by sponsor Hallmark; he was, for example, able to give his cast fifteen days of rehearsal for a ninety-minute program, leading to a professional patina rare on any stage.

Though Schaefer worked in film and theater as well as television, he preferred the small screen, because it combined the theatricality of the stage with the directorial control of the screen. For teletheater, the director had great control over what the viewer at home saw and heard: "Part of the audience can hate the scene, part can like it, some can even scorn it or laugh at it; as long as the scene's exciting and involving the audience, it's successful in television terms." In the Golden Age of television drama, the Schaefer years were twenty-four karat.

George Schaefer with Maurice Evans and Judith Anderson during rehearsal for Macbeth, *"Hallmark Hall of Fame," 1960.*

Photograph by Declann Haun. George Schaefer

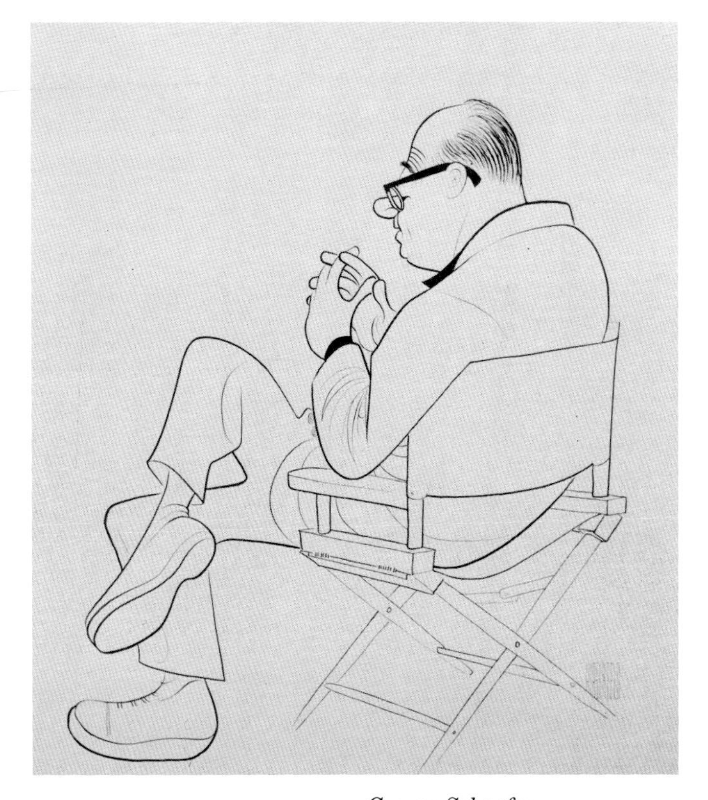

George Schaefer

India ink on artist board by Al Hirschfeld. George Schaefer

Paddy Chayefsky 1923–1981

Paddy Chayefsky specialized in intimate television drama—a genre that he helped create and shape into its finest form. The medium proved an amenable arena for what critic Arthur Knight has typified as dramatic "chamber works"—plays stressing the everyday and commonplace rather than the grand or epic. The small screen gave precise focus to introspective, internalized theater. Chayefsky said, without regret, in 1956 that "I have sometimes been accused of writing little plays about little people. What my critics pretend to mean, I think, is that my plays are literal and earthbound, and that my characters never achieve any stature beyond immediate recognition."

Paddy Chayefsky, 1954.

Wisconsin Center for Film and Theater Research

Rod Steiger and Nancy Marchant in Paddy Chayefsky's Marty, *performed on "Goodyear Playhouse," May 24, 1953.*

Photograph by Gary Wagner. Wisconsin Center for Film and Theater Research

Chayefsky wrote essentially in the vernacular, and nowhere was this more successful than in his 1953 television play, *Marty. Marty* was a small-focus slice of the mundane, centered on what Chayefsky called "the most ordinary love story in the world" between a shy, homely Bronx butcher, Marty (Rod Steiger), and the plain-looking Clare (Nancy Marchand); incidental characters included Marty's mother (Esther Minicotti) and his best friend, Angie (Joseph Mantrell). What operated beneath this ordinary love story was Chayefsky's theme—that somehow in the second half of the twentieth century people have to find a way not to be crushed by historical tides, which "are too broad now to provide individual people with any meaning to their lives. People are beginning to turn into themselves, looking for personal happiness. . . . The jargon of introspection has become everyday conversation." In *Marty*, Chayefsky set out to rescue the butcher—or to contrive to let Marty rescue himself—from a life that had become so overwhelmingly minimalist as to be microscopic. He worked this by making it a love story, and, however commonplace, Marty's love for Clare ennobled them both. It was a touching, comprehensible resolution.

The most enduring line of the play—and perhaps of all Golden Age television drama—was the mantra Marty used to mumble to his pal Angie. What line could better encapsulate the not-larger-than-life, the utterly mundane, the oppressively domestic vernacular than when Marty routinely says, "I don't know, Angie. What do you feel like doing?"

Rod Serling 1924–1975

Called television's "only angry man" in the mid-1950s, Rod Serling started as a radio and television scriptwriter and a modestly successful freelance dramatist. His breakthrough came in 1955, when *Patterns*—his play boring in on the top echelons of corporate America—was telecast on the "Kraft Theatre." Hailed as "one of the high points in the TV medium's evolution," *Patterns* established Serling as the most in-demand writer in the business. In 1955 alone, twenty of his plays were produced for television.

Serling's best-remembered play, however, was *Requiem for a Heavyweight*, which inaugurated the "Playhouse 90" series on October 11, 1956. This drama about a broken-down prizefighter remains one of the high points of television's Golden Age. *Requiem*, raved one critic (a trifle condescendingly), proved that "the plays of this little electronic theatre . . . stand up with some of the best dramatic writing of our time."

In the late 1950s Serling gradually stopped writing live television drama. The direction of the industry was not encouraging: television production was rapidly changing, and not in a manner suited to live drama. More and more, the locus of the entire industry moved from the East Coast to Hollywood, and legions of writers from the ranks of B movies supplanted the nucleus of writers who had grown up in radio, television, and the theater. Serling switched from drama to science fiction with the series "The Twilight Zone." Although he had had more than one hundred of his plays telecast by 1960, he wearied of the kinds of compromises that had become necessary to produce live television drama: "I simply got tired of battling . . . a sponsor, a pressure group, a network censor. . . . It's a crime, but scripts with a social significance simply can't be done on TV." It was, perhaps, one mark of television's coming of age. Its very ubiquity—and the corresponding size of its audience—vitiated against controversy. Serling figured, and rightfully so, that science fiction and fantasy were far enough removed from reality to be successful.

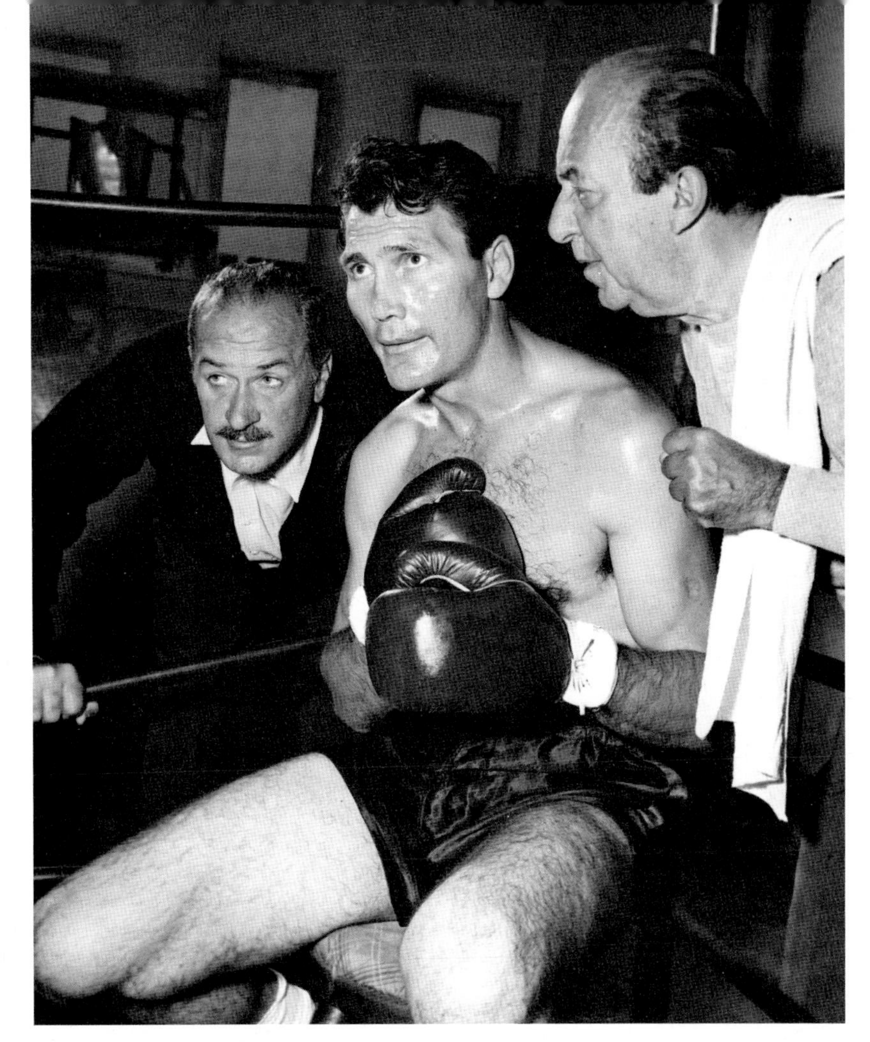

Keenan Wynn, Jack Palance, and Ed Wynn in a scene from Requiem for a Heavyweight, *broadcast on "Playhouse 90," October 11, 1956.*

CBS Inc.

Rod Serling

Museum of Broadcasting

Sitcoms and Serials

One of the program formats that broadcasting borrowed from the print medium was the serial, a popular nineteenth-century form of narrative fiction written in installments for mass-circulation magazines. The serial formula featured a regular cast of characters (often multigenerational) and a daily or weekly plot that was generally (but not always) continuous. On radio and television, serials took shape as a phantasmagoria of domestic dramas, situation comedies, and adventure series cast in Western, detective, or mystery modes. Over and over again, the moral message of the serials summoned us to stay tuned—to truth, justice, and the American way.

Lucy

Broadcast Pioneers Library

"Amos 'n' Andy"
Freeman Gosden (Amos)
1899–1982
Charles Correll (Andy)
1890–1972

Freeman Gosden (Amos) and Charles Correll (Andy), circa 1928.

NBC

Every night at 7:00, as an organ played the strains of "The Perfect Song," soft-voiced announcer Bill Hay would say, "Here they ah, Amos 'n' Andy." For the next fifteen minutes the business of the nation came to a halt—with dinners, mah-jongg, movies interrupted—while Americans gathered around their radios for a visit with the proprietors of the Fresh Air Taxicab Company, Incorporated. Costarring with Amos 'n' Andy in this broad comedy serial were Madame Queen, Kingfish George Stevens, Sapphire, Lightnin', and the Mystic Knights of the Sea. "Amos 'n' Andy" became a nightly ritual during the Depression, with an estimated one-third of the nation as regular listeners. Restaurants, bars, even movie theaters bought radio sets and rolled them into listening range every night from 7:00 to 7:15; it was the only way to lure customers from their home sets. The escapades of Amos Jones (Freeman Gosden) and Andrew H. Brown (Charles Correll) became topics of national conversation: the country rallied to Andy's defense in a riotous breach-of-promise suit, and millions of listeners suggested names for Amos's baby girl. And often as not the national conversation itself was sprinkled with buzzwords like "Awa, awa, awa," "Um-um, ain't dat sumpin?" and "Don't get me regusted."

This most popular of all radio programs began with two white men who first teamed up to perform in minstrel shows in 1919. Moving to Chicago, they had a late-night half-hour show in 1925 on WEBH, in the Edgewater Beach Hotel. They were such a hit that the *Chicago Tribune*-owned station, WGN, asked them to develop a comedy serial for radio based on "The Gumps," a comic strip in the *Tribune*. Instead, Correll and Gosden created a serial about two blacks named Sam and Henry. Premiering in January 1926, "Sam and Henry" was an instant success. Two years later WMAQ stole them away—but, as the *Tribune* claimed rights to the title "Sam and

Andy counting the receipts of the Fresh Air Taxicab Company while Amos looks on, "regusted," April 10, 1930.

Library of Congress

Henry," the name of the show was changed to "Amos 'n' Andy."

What happened in the next year transformed radio. The show became such a phenomenon that it is credited with popularizing radio itself, with changing the perception of radio from that of interesting toy to institution, making it a regular part of everyday life—much as David Sarnoff had prophesied in his "Music Box Memo" of 1916. Sponsors reevaluated their attitudes about radio, watching as Pepsodent—which had taken a chance by sponsoring the program on a nationwide NBC hookup—benefited from the show's resounding success. It was obvious that radio could be a lucrative market, and commercialization of the medium rolled forward with a vengeance.

Relying heavily on burnt-cork dialect, Correll and Gosden portrayed two black men who had arrived in Harlem from rural Georgia during the black migration from "field to factory" in the 1920s. Intent on making their fortune, they get cajoled into buying an old rattletrap car from the Kingfish to start up a taxicab company. The car was such a wreck that it lacked even a top—ergo, the "Fresh Air Taxicab Company." Correll once contrasted Amos and Andy's characters: "Amos is a hardworking little fellow who tries to do everything he can to help others and to make himself progress, while his friend Andy is not especially fond of hard work and often has Amos assist him in his own duties." To say the least.

Correll also suggested that one reason for the show's addictive power was that the characters were likable: with daily exposure to the radio audience, they built up a solid fund of fondness. The program in fact focused primarily on characters rather than gags, with the humor deriving from human foibles—from Andy as the prototypic shiftless but endearing relative to the Kingfish's comic knavery. They were characters grown in minstrels and vaudeville but with roots in Dickens (e.g., *Nicholas Nickleby*) and Twain (as in the Duke in *Huckleberry Finn*).

For several years Gosden and Correll wrote all of their own scripts, about two thousand words a day, and went straight to broadcast without rehearsal. In addition to Amos, Gosden played the Kingfish and Lightnin'; Correll played Andy, Henry van Porter, and Brother Crawford. Both also played scores of other characters.

"Amos 'n' Andy" precipitated a flood of ethnic and dialect serials—shows like "Lum and Abner," "Life with Luigi," and "Abie's Irish Rose." Ethnic humor propelled radio's popularity: the irony is that radio's relish for diversity paved the way for the emergence of a national popular culture in the 1930s.

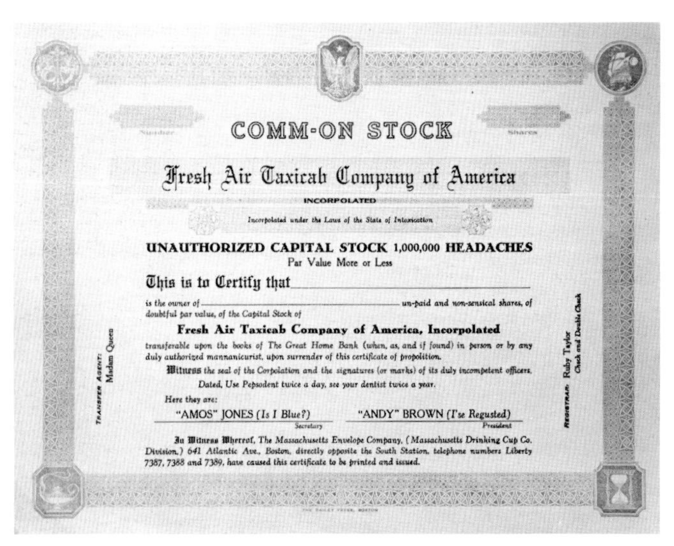

COMM-ON STOCK

Fresh Air Taxicab Company of America

INCORPOLATED

Incorpolated under the Laws of the State of Intoxication

UNAUTHORIZED CAPITAL STOCK 1,000,000 HEADACHES

Par Value More or Less

This is to Certify that ______

is the owner of ______ un-paid and non-sensical shares, of doubtful par value, of the Capital Stock of

Fresh Air Taxicab Company of America, Incorpolated

transferable upon the books of The Great Home Bank (when, as, and if found) in person or by any duly authorized mannanicurist, upon surrender of this certificate of propolition.

Witness the seal of the Corpolation and the signatures (or marks) of its duly incompetent officers.

Dated, Use Pepsodent twice a day, see your dentist twice a year.

Here they are:

"AMOS" JONES (*Is I Blue?*) — *Secretary*

"ANDY" BROWN (*I'se Regusted*) — *President*

TRANSFER AGENT: Madam Queen

REGISTRAR: Ruby Taylor, Check and Double Check

In Witness Whereof, The Massachusetts Envelope Company, (Massachusetts Drinking Cup Co. Division,) 641 Atlantic Ave., Boston, directly opposite the South Station, telephone numbers Liberty 7387, 7388 and 7389, have caused this certificate to be printed and issued.

Comm-on Stock/Fresh Air Taxicab Company of America.

Broadcast Pioneers Library

Gertrude Berg 1899–1966

"The Rise of the Goldbergs" was, from 1929 to 1950, one of the best-known ethnic soap operas on radio. Produced, written, and acted by Gertrude Berg, the program—later shortened to "The Goldbergs"—was a continuing series of Lower East Side vignettes, with a heavy reliance on domestic drama and likable Yiddish characters. The program began with Molly (Gertrude) yelling up the airshaft, "Yoo-hoo! Mrs. Bloo-oom!" The other main characters were Molly's husband, Jake Goldberg, and their children, Sammy and Rosie, who "helped teach their immigrant parents how to become Americans." The parents, in turn, "tried to teach them some of the rich traditions of the Old World."

Molly was the quintessential do-gooder. When Jake asked her, "Vhat tanks vill you get? Running your feet off far your neighbors, ha?" She replied, "To do good things, dat ain't itself a tanks? Notting no matter vhat you vould do is so sveet like vhen you do something far somebody."

Berg wrote the scripts for this six-day-a-week, fifteen-minute show by carefully mapping out the story sequence: she would choose which character would be central and block out the plot, first by weekly and then by daily episode. "The Goldbergs"—sponsored originally by Pepsodent, next by Palmolive, and then by Procter and Gamble (inspiring the "soap" appellation)—insisted on as much sudsy realism as possible. If the telephone was supposed to ring, a real one was made to ring; if the script called for fried eggs, real eggs were broken into a real frying pan and sizzled in front of the microphone.

Gertrude Berg had an intuitive feeling for the subtlety of unseen drama and did much to create and to popularize family serial strips on radio. She based her characters on real people and put them in credible situations. It all worked, she once said, because "People are most interested in each other."

Gertrude Berg, creator, writer, and star of "The Goldbergs," poses with ten years of scripts that she wrote for the show.

Photograph by George Karger. *LIFE* magazine © 1949 Time, Inc.

Paul Rhymer 1905–1964

In the 1930s and 1940s on NBC, radio scriptwriter Paul Rhymer brought viewers a glimpse inside "the small house half way up the next block" in his daytime serial, "Vic and Sade." Here was small-town life at its winning best, gently focused on the everyday; it was, someone once said, "like eavesdropping on friendly next-door neighbors." Rhymer began as a staff writer for NBC in 1930 in Chicago—in those days the proving ground for radio drama—and there developed the saga of Victor Rodney Cook and his wife Sade, their twelve-year-old son Rush, and his barkless dog, Mr. Albert Johnson.

TIME magazine called Rhymer's writing "gentle, funny, low-key, and as timeless as the telephone poles on U.S. 20." Writer Jean Shepherd, who grew up in Indiana listening to "Vic and Sade," has said that, "As far as I know, no one working in the mass media has ever created such a complete and flawless world, peopled with characters so fully realized." As a daytime serial, "Vic and Sade" shared the air with a banal band of soaps—a fate, Shepherd also said, comparable to having *Death of a Salesman* or *Our Town* debut on a Wednesday afternoon following "As the World Turns."

Rhymer's voice was more ironic than sentimental, closer to Garrison Keillor than to Thornton Wilder. Edgar Lee Masters—with whose works Rhymer bore more than passing acquaintance—called the serial the best of its time. We learned about Fred and Ruthie Stembottom's snail-like trip to Chenoa, Illinois, Ike Kneesuffer's indoor horseshoe set, Uncle Fletcher's wandering mind, and

Paul Rhymer

State Historical Society of Wisconsin

Vic (Art Van Harvey) and Sade (Bernardine Flynn) at microphone.

Wisconsin Center for Film and Theater Research

Vic's lodge brothers in the Sacred Stars of the Milky Way. So evocative was the way in which these characters were woven into the script that listeners barely realized that most of them never appeared before the mike: they were purely creatures of Vic and Sade's conversations. Dialogue dominated story in "Vic and Sade," with each episode complete in and of itself. Often they revolved on a kind of mundane absurdity, as in the show "The Washing Machine Is on the Blink." Here the entire episode delineated how the deus-ex-washing-machine conspired against do-it-yourselfery in a statement about the guaranteed breakdown of modern technology.

Rhymer had a wonderful ear for the way in which people actually communicated, and "Vic and Sade" never lost this connection with its audience. It was radio daytime drama at its peak.

Carlton E. Morse born 1901

The creator of "One Man's Family," Carlton E. Morse was radio's most successful serial dramatist. From 1932 to 1959 he narrated the story of three generations of the Barbour family of Sea Cliff, California.

Morse draws a strong distinction between "One Man's Family" and run-of-the-mill soaps. "For soap operas, they think of a plot and then drop people in to tell the plot. In my show, first came the characters. Their relationship to each other is what made the plot." His original inspiration for writing radio drama came from John Galsworthy's novel *The Forsyte Saga*. "I thought, why wouldn't a good family story be of interest [on radio]?"

He was right, and when "One Man's Family" went on the air on April 29,

Carlton E. Morse, creator of "One Man's Family."

Broadcast Pioneers Library

Listener's Choice-Radio Guide Award for "One Man's Family," 1940.

Carlton E. Morse and Trustees, Morse Family Trust

1932, it found a home for the next twenty-seven years. Instead of rewriting stage shows for radio, Morse wrote directly for the medium, discovering that "To write just for the ears turned out to be a special technique."

He also wrote the long-running series "I Love a Mystery" (1939–1952) and forty-three other serials, including "The Woman in My House," "Family Skeleton," and "Adventures by Morse." Between 1939 and 1945, with both "Family" and "Mystery" running daily, Morse would be at his typewriter at dawn and would stay there until 9:30 P.M. He estimates that he ground out ten million words for "Family," three for "Mystery," and another million for assorted other shows.

But "One Man's Family" remained Morse's greatest legacy. Over the years the program became a national ritual. Millions of Americans grew up, married, and got old with the program's characters. Morse has talked of the "feeling of reality that no other show could duplicate. *One Man's Family* moved with the slowness of life itself, working on tiny pieces of characterization and subtle, underlying contact." The program would open with the announcer dedicating each episode "to the mothers and fathers of the younger generation and to their bewildered offspring." Then he would announce the episode's book and chapter; the last broadcast, on May 8, 1959, was Chapter 30 of Book 134. More than one hundred characters appeared over the years, with the action centered around father Henry Barbour, his wife Fanny, and their five children—Paul, Hazel, twins Clifford and Claudia, and Jack. For 3,256 episodes, Morse made the Barbour family America's own.

"Fibber McGee and Molly"
Jim Jordan 1896–1988
Marian Jordan 1898–1961

When Fibber McGee reached for the door to his hall closet at 79 Wistful Vista, millions of Americans tensed and automatically yelled, "NO—DON'T!" Ephemera of the 1930s and 1940s filled that closet—the moose head carefully put away in 1936, the anchor bought for the boat he never built, the mah-jongg board, the pairs of outgrown ice skates. "Tain't funny, McGee," came the voice of the infinitely patient, amiable Molly, as the accumulated clutter crashed to McGee's feet. The one time the closet had performed a useful function was the day when burglars broke in, tied up McGee, and demanded to know where the family valuables were kept. Of course. The bandits were still under the rubble when the police arrived.

Jim and Marian Jordan had spent the first several years of their married life as one of the reputedly worst vaudeville teams ever to tour Midwestern tank towns, opera houses, and churches. They would stand in front of a velvet drop cloth emblazoned with their names and sing good-natured songs like "When You're Smiling" and "Side by Side." Eventually they went broke.

But radio beckoned, and after breaking into broadcasting in the 1920s and early 1930s, they emerged as "Fibber McGee and Molly" over network radio in 1935. The rest of the cast included the Great Gildersleeve (Harold Peary), Mayor La Trivia (Gale Gordon), and Wallace Wimple (Bill Thompson), who played the supremely henpecked husband married to "my big old wife, Sweetie-face." Fibber was forever trying to invent something useful like "self-peeling bananas," and Molly would gently support his enthusiasms, usually encouraging him with the expression "Heavenly days."

Fibber has been called "the Everyman of the Depression." Even their address, on "Wistful Vista," evoked a hopeful small-town nostalgia. Along with Jack Benny's bank vault, Duffy's Tavern, and Allen's Alley, the McGees and their "dad-ratted" closet became part of American folklore.

Fibber McGee and Molly

From *Great Radio Personalities in Historic Photographs* by Anthony Slide, Vestal Press, 1988

"Easy Aces"
Goodman Ace 1899–1982
Jane Ace 1905–1974

The "Easy Aces" was a seminal radio sitcom written, produced, and acted by the urbane Goodman Ace and his apparently addlepated wife Jane. First on network radio in 1931, the program featured dazzling dialogue and a meandering, seemingly nondirected plot. Though the words were Goodman Ace's, the program's star was really Jane, the very model of a modern Mrs. Malaprop. Among her utterances were "reasons too humorous to mention," "he's a ragged individualist," and her description of her husband as "a human domino."

Ace, whom Fred Allen called "America's greatest wit," reveled in radio's imaginative power: "In radio, the best thing you had going for you was that little one-inch screen inside your brain, and when the mental picture flashed, you saw what you wanted to see." He cast himself in the "Easy Aces" as an advertising man in "the typical little Eastern town known as New York City." Other characters included his boss—renowned for having a cliché for every occasion—and Jane's brother Paul, who had not worked for twelve years, as he was waiting for the dollar to stabilize.

Sometimes the plots stopped in mid-program, rearranged themselves, and went off in entirely new directions, probably to suit a new bit of dialogue. The audience didn't seem to mind. Ace said, "I learned a long time ago from Groucho Marx: 'Don't complain and don't explain.' People don't listen to what you say anyway. It's all just noise."

But what delicious noise the "Easy Aces" broadcast until 1945. Their polished sitcom was one of the comedic high points of radio's Golden Age, but became bogged down after the war in the popular craving for something new and different. People wanted new formats and were tired of programming that had grown ritualized—not to say stale—in the two previous decades. And, while radio floundered around looking for new solutions, television—the shiny new toy—sneaked in with old vaudevillians and stole the show.

The "Easy Aces"— Goodman and Jane Ace.

From *Great Radio Personalities in Historic Photographs* by Anthony Slide, Vestal Press, 1988

Scene from "I Love Lucy"—the "Chocolate Factory" episode.

CBS Inc.

Lucille Ball born 1911

In the Golden Age of television, Lucille Ball wore comedy's crown firmly on her red head. The irrepressible comedienne was characterized as "the distaff equivalent of Jack Benny" and, because of her skill at pantomimed slapstick, as that rarest blend of comic artistry—"the clown with glamour." Husband Desi Arnaz was a Cuban bandleader who pioneered innovative TV production methods, especially in the use of film.

The two met on the movie set of *Too Many Girls* and married in 1940. They broke into broadcasting a decade later, when Lucy starred in the radio sitcom "My Favorite Husband." They organized Desilu Productions in 1950 to coordinate their business and performance schedules, and the following year filmed the first "I Love Lucy." Basically, each episode was filmed as if it were a three-act play: three cameras were used, and the show was performed before a live studio audience. After editing and scoring, it was then released for television. The use of film was a landmark in small-screen production, light-years removed from the evanescence of live TV and affording the prospect of countless years in syndication. In fact, the financial success that redounded to Desilu radically transformed television production: what had been a medium that was 80 percent live was by the mid-1950s pointed solidly toward film. Hollywood writers, directors, actors, and production crews flocked to television, and the New York television production base shifted to the West Coast. The advent of videotape in 1956–1957 only speeded the upheaval.

CBS was the host network for "Lucy's" premiere on October 15, 1951. Miss Ball's costar was real-life husband Desi Arnaz; though William Paley had at first objected to Desi's heavy Cuban accent, Lucy had insisted, conga drums and all.

The supporting cast—Vivian Vance as Ethel and William Frawley as Fred—were everybody's aunt and uncle; could there ever have been a time when we didn't know these people? Certainly "I Love Lucy" became the godmother of all TV sitcoms—which is not to blame it for the eventual degeneration of the species.

Lucy and Desi played enjoyable characters in implausible situations: Desi as the bandleader "Ricky Ricardo" who worked at the Tropicana, Lucy as the housewife who hungered for a career in show biz. The first episode was "The Girls Want to Go to a Night-Club," and its plot hinged on Ricky and Fred's subterfuge to get to go to the fights; meanwhile, Lucy and Ethel had arranged for them all to go out clubbing to celebrate the Mertzes' anniversary. The resulting chaos may not have been Eugene O'Neill, but it was certainly funnier.

The series' best-known episodes were those surrounding the birth of the Ricardo (and Arnaz-Ball) baby. The show of January 19, 1953—"Lucy Goes to the Hospital"—garnered what was then the largest viewing audience ever (69 percent of all sets), and it far overshadowed the inaugural festivities for Dwight Eisenhower. There were scores of other memorable Lucy episodes—like "Lucy's Italian Movie," in which director "Vittorio Fellipi" gives her a role in his new movie *Bitter Grapes*, and she winds up slinging grapes at her Italian co-worker in the stomping vat; or the day when Lucy and Ethel audition for jobs as candy workers at Kramer's Candy Kitchen and lose to a berserk conveyor belt; or the time when Lucy does a TV commercial and gets schnockered by drinking too much "Vitameatavegamin," which is 23 percent alcohol.

Desi once said that the reason "I Love Lucy" worked was that "the chemistry, the mixture of the people there, made it fun." What they made it was family.

Lucy and Ricky (Desi Arnaz) with Ethel (Vivian Vance) and Fred (William Frawley), November 1956.

CBS Inc.

Scene from "Dragnet," January 22, 1952, with Sergeant Joe Friday (Jack Webb) and Officer Frank Smith (Ben Alexander).

NBC

Jack Webb 1920–1982

On radio in the late 1940s, Jack Webb developed the laconic, tough-guy persona that emerged full-blown as Sergeant Joe Friday in "Dragnet." To avoid yet another shoot-'em-up cop melodrama, Webb asked the Los Angeles police chief for advice about real police work; the chief told him not to make heroes of the cops and not to avoid depicting the day-to-day drudgery. Webb agreed, and he gave the series a low-keyed, almost documentary atmosphere.

"Dragnet" began as a radio show in 1949 and premiered as a regular series on NBC television in January 1952. Webb not only drew from the Los Angeles police files for cases but submitted scripts to the police department for technical accuracy.

Jack Webb

Wisconsin Center for Film and Theater Research

Public response was surprising: although the vogue for mysteries and detective serials had never really waned on radio in the 1940s, "Dragnet" managed to challenge "I Love Lucy" as television's number-one program. By refusing to use the clichés of traditional crime shows, Webb created a whole new genre of realistic, mystery-crime serials. Other crime shows done in this television-*noir* style soon followed, including "Big Town," "Highway Patrol," "The Lineup," and "Racket Squad."

Webb's stone-faced portrayal of Sergeant Friday was originally teamed with Ben Alexander's Officer Frank Smith. The "Dragnet" set was a reconstruction of the lower floor of the Los Angeles Police Department, and the documentary style—clipped and spare—was reinforced by the "Just the facts, Ma'am" dialogue and extensive use of close-ups.

The series was a major television influence in the early and mid-1950s, striking an enormously responsive chord from its opening four-note theme (dum-de-dum-dum) to the hammered clank of the "Mark VII" trademark at the end. Webb's characterization of Joe Friday achieved the ultimate form of flattery by becoming one of the most parodied personalities ever to appear on television.

Raymond Burr born 1917

Adventure series came as naturally to television as they had to radio, with such genres as mysteries, westerns, or some permutation of urban cowboy thrillers dominating the dial by the mid- and late 1950s.

Erle Stanley Gardner, creator of the fictional defense attorney Perry Mason, had been reluctant to try television, but when he saw Raymond Burr's video test, he yelled, "That's him! That's Perry!" Burr came to the role after a decade in Hollywood playing "heavies," notably as the henpecked murderer in Alfred Hitchcock's *Rear Window*. In "Perry Mason" Burr effectively used his size—6 feet 2 1/2 inches and more than 210 pounds—to project a looming courtroom presence.

The show debuted on CBS in September 1957, opposite Perry Como. Critics were less than adulatory, one describing Burr's Mason as too much a lusterless "resident of suburbia." But the television audience itself soon discovered a new hero in Burr's quietly virtuous depiction of Perry Mason. If not quite perceived as an endearing teddy-bear type, he did come across as likable and totally forthright. With a strong supporting cast, which included Bill Hopper as Paul Drake and Barbara Hale as secretary Della Street, Burr conducted his weekly courtroom duels in the Manichaean fashion of the 1950s, Good versus Evil, plain and simple. The 30.5 million people who watched "Perry Mason" never had to worry about who was going to win.

Raymond Burr

CBS Inc.

James Arness born 1923

When "Gunsmoke" was first developed by CBS radio in 1952, it began with three major premises: first, this was not to be a horse opera but a serial involving realistic characters; second, no one would ever wear spangles and fringe and say things like "sidewinding varmint"; and third, there would be no horse costar, much less one that did tricks.

Veteran actor William Conrad took the radio role of Matt Dillon, but it was James Arness who brought the character to television in 1955. Arness's portrayal of the Dodge City marshal was of a piece with Owen Wistar's Virginian or Hollywood's John Wayne: chiseled in a kind of towering granite perpetuity.

Along with Arness's lantern-jawed presence came a superb cast—Amanda Blake as Kitty, Milburn Stone as Doc, and Dennis Weaver as Chester. "Gunsmoke" was in the vanguard of the craze for "adult westerns" and in fact helped to spark it. An instant hit, the show was first in ratings in 1957 and among the top ten thereafter.

Because "Gunsmoke" was written as an anthology, with the focus shifting from character to character on different episodes, Arness didn't really overpower the others. The ensemble was warm and credible, and together they imbued the series with a rough-hewn solidity and an unbending sense of good-guys-versus-bad. As Matt Dillon said, "Everybody's got a right to live in this town if they mind their own business and stay within the law."

"Gunsmoke" cast: Milburn Stone, Amanda Blake, James Arness, and Dennis Weaver, May 7, 1958.

CBS Inc.

Alfred Hitchcock 1899–1980

Alfred Hitchcock, the master of suspense, produced television's most successful psychological series in the 1950s, "Alfred Hitchcock Presents." Unlike many of his Hollywood colleagues, Hitchcock—though he bore the small screen no great love—recognized the medium's commercial possibilities.

Hitchcock's brilliant films, such as *The 39 Steps, Rebecca, Suspicion, Spellbound, Notorious, Rope, Dial M for Murder, Rear Window*, and *Vertigo*, relied heavily on leading audiences down the path of the unexpected. His direction created mounting rhythms of anticipation and bursts of terror, supplemented by a device that he called a "MacGuffin." A MacGuffin was whatever trick or hook he might invent for each movie to serve as a nagging leitmotif, whether it be a recurring snatch of music or the continued appearance of a character with some strange quirk.

In 1955 Hitchcock's agent, Lew Wasserman, head of MCA (parent company of Universal Studios), persuaded him to get into television production. Hitchcock's decision to go ahead would make him one of TV's most recognizable personalities; for, though he directed only a few of the show's episodes, he contributed regularly to script selections. And of course there were his introductions. He would step into his famous self-caricature as the strains of his theme song, "March of the Marionettes," faded. Then he would deliver his deadpan greeting, "Good evening." The public especially enjoyed his jabs at sponsors: he said of commercials, "Most are deadly. They are perfect for my type of show."

Hitchcock employed first-rate writers (including Ray Bradbury), good actors, and strong scripts. The standards of production were comparable to those of such live-drama shows as "Playhouse 90" and "Hallmark Hall of Fame." But his success also rested on the lighthearted way in which he treated macabre situations. A master stylist, this genial ghoul once said, "In selecting the stories for my television shows, I try to make them as meaty as the sponsor and the network will stand for."

The first episode he directed himself was "Breakdown," starring Joseph Cotton. Cotton played a businessman who has an apparently fatal car accident. Thrown into a mortuary, he is about to be embalmed when a tear appears in his eye, and he is saved—just—from the grotesque terror of being buried alive.

As he did with his films, Hitchcock used humor to relieve horror in his TV series, believing that "after a certain amount of suspense, the audience must find relief in laughter." American audiences found a great deal of suspense and laughter on "Alfred Hitchcock Presents" and made Hitchcock himself a virtual cult figure.

Alfred Hitchcock

Gouache by Al Hirschfeld. The Margo Feiden Galleries, New York City

Children's Shows

Although television fueled children's programming with a commercial zeal at which radio—with "premiums" and "secret-decoder rings"—had only hinted, the child-audience was not suddenly discovered on television's doorstep. Network radio had developed a devoted juvenile following after 1930, both with comic-strip shows like "Little Orphan Annie" and "Jack Armstrong, All-American Boy," and with quality programs such as Nila Mack's "Let's Pretend" and Ireene Wicker's "Singing Story Lady." The latter type of programming particularly met the children's code that NBC and CBS adopted in the mid-1930s.

In establishing a code that treated radio as virtually a socializing agent for America's youth, NBC announced that "All stories must reflect respect for law and order, adult authority, good morals and clean living. The hero and heroine, and other sympathetic characters, must be portrayed as intelligent and morally courageous. The theme must stress the importance of mutual respect." CBS declared that "the exalting, as modern heroes, of gangsters, criminals, and racketeers will not be glorified or encouraged. Programs that arouse harmful nervous reactions in the child must not be presented. . . . Dishonesty and deceit are not to be made appealing or attractive to the child."

One of radio's best attributes—its power to touch the imagination—had been used to great effect with children's shows. With television, puppets, cartoons, and stylized adventure heroes helped to bridge the imaginative gap between fantasy and reality, and soon these became as much a part of childhood as children's radio or Saturday matinees at the Bijou had been before. What television discovered was the lucrative commercial market that children afforded. TV's visual dimension, allowing kids to *see* Howdy Doody or the Mickey Mouse Club or Davy Crockett, created an ever-expanding demand for related goods. Here, as with adult game shows and amateur hours, broadcasting hinged its ratings success to audience participation: watching the Mickey Mouse Club sans the "ears" was tantamount to missing part of the experience. It was as bad as not knowing all the words to the club's theme song, and it meant, quite simply, not sharing in full membership. The children's market proved to be one of television's great mother lodes, for what child in America wanted to be excluded from the club?

Buffalo Bob, Clarabell, and Howdy Doody.

NBC

Nila Mack circa 1895–1953

Every Saturday morning for nearly twenty-five years, Nila Mack wrote, directed, and produced one of radio's finest children's programs, "Let's Pretend." She came to radio in the late 1920s after an acting career that included six years with Alla Nazimova's film company and extensive touring in vaudeville and on Broadway. She first worked for CBS as an actress in the "Radio Guild" productions, a series that formed the basis for the network's highly acclaimed sustaining program, "Columbia Workshop." In 1930 the network asked her to direct its children's show, "The Adventures of Helen & Mary," which, as "Let's Pretend," eventually became CBS Radio Network's oldest continuous dramatic series.

Nila Mack, longtime host of CBS children's program "Let's Pretend."

CBS Inc.

In a 1952 interview, Miss Mack recalled that children's entertainment was at a low ebb when she began "Let's Pretend" in the Depression: "I remembered fairy stories that filled me with wonder when I was very young. I figured that if these lively pieces with a message at their hearts had meant so much to me, other children would like them too." She adapted such classics as the stories of Hans Christian Andersen, the Brothers Grimm, and the *Arabian Nights*.

Like another highly popular radio show for children—NBC's "The Singing Story Lady" with Ireene Wicker—"Let's Pretend" focused on storytelling and was filled with "kings and queens and princes who ride talking horses through enchanted forests." Miss Mack hired mainly children for her casts, believing that they could best convey the sense of fantasy and magic that she wanted transmitted to her audience. A complete tale was told each week, and there was never any dithering about a story's outcome. "The good are very good," she explained, "and the bad get just what they deserve." The program's sound effects were especially wonderful, as when terrific sizzling and steaming noises were produced to indicate boiling oil being poured over Ali Baba's forty thieves. Usually, Miss Mack preferred the power of suggestion over graphic depiction, as in the story of Bluebeard's fall: "The kids are tickled to death when Bluebeard's sword falls, klunk, closely followed by the thud of Bluebeard hitting the ground for the last time. They get the idea, but not the horror." Classic heroes always held much more interest for her than newer radio heroes like Buck Rogers or Jack Armstrong. "I'll back seven-league boots and magic wands any time," she once said, "against six-shooters and spaceships."

"Let's Pretend" won radio's most distinguished accolades over the years, including the Peabody Award and recognition as the "best children's program in radio."

"Kukla, Fran, and Ollie"
Fran Allison born circa 1907
Burr Tillstrom 1917–1985

Yes, by gum and yes by golly
Kukla, Fran and dear old Ollie.

The Kuklapolitans were pure television magic. More than puppets, they were a repertory company sprung from the creative rib of Burr Tillstrom. Starring the bald, bulbous-nosed Kukla (Russian for "doll") and snaggletoothed but gentle Oliver J. Dragon, the program had a supporting cast that included Olivia Dragon (Ollie's mother), Fletcher Rabbit (the droopy-eared mailman), Ophelia Oglepuss (former opera diva), and Beulah Witch (who patrolled the network's coaxial cable on her jet-powered broomstick). The sole nonpuppet on camera was Fran Allison, a radio star from the "Don McNeill Breakfast Club"; Burr Tillstrom made all of the puppets and did all of the voices.

"Kukla, Fran, and Ollie" began in postwar Chicago, then a hotbed of television experimentation. "We made up television," Tillstrom once said. "There was no influence to teach us." Beginning in 1947 on WBKB in Chicago, the show was picked up a year later by NBC, and it continued to captivate children and adults for ten years. The ensemble worked together so closely that they seemed cut from the same cloth, all somehow real. Fran Allison has said, "We found we had many interests in common," and "We kind of grew up together." The show was done live and without a script, relying on inventive adlibbing from all and a general sense of camaraderie and affection. Tillstrom once explained, "If I ever plotted exactly what I was going to do, Kukla and Ollie wouldn't work for me. They don't like cut-and-dried stuff. Fran wouldn't be able to talk to them, either." Occasionally they would do special features on the program—"Martin Dragon, Private Tooth," or the *Mikado*, with Kukla as Nanki Poo, Fran as Yum-Yum, and Ollie as the Lord High Executioner. Arthur Fiedler and the Boston Pops once accompanied a skit called "St. George and the Dragon."

Burr Tillstrom and Fran Allison with puppets Kukla and Ollie.

Chicago Historical Society

On the historic 1953 "Ford 50th Anniversary Show"—a spectacular broadcast done simultaneously on NBC and CBS and featuring every big star in the galaxy, from Marian Anderson to Edward R. Murrow—"Kukla, Fran, and Ollie" offered some gentle banter. Appearing just after Ethel Merman's rendition of "There's No Business Like Show Business," Ollie said, "There's no business like television, either." Kukla responded, "It's so *young*, you know." To which Ollie replied, "Yes—when will it grow up?"

Whether or not it ever grew up, it never again equaled the creative warmth of Tillstrom's troupe. The Kuklapolitans showed us how good television could be. "I don't think we ever intended it for kids," Tillstrom once said. "Not for them alone, at least. We've always assumed that *this* family was for the whole family."

Howdy Doody

It's Howdy Doody time, it's Howdy Doody time,
Bob Smith and Howdy too, sing howdy do to you.
Let's give a rousing cheer, 'cause Howdy Doody's here,
It's time to start the show, so kids let's go!

One of the earliest TV shows for kids, "Howdy Doody" aimed straight for the Peanut Gallery. Howdy himself was a marionette invented by NBC's puppet-maker Velma Dawson and a Buffalo, New York, disc jockey named Bob Smith. Red-haired, freckled, and blue-eyed, Howdy was theoretically a ten-year-old boy, and was decked out in jeans, flannel shirt, and cowboy hat and boots. Doodyville's other residents included Clarabell, a mute, seltzer-wielding clown; Phineas T. Bluster, a crusty old geezer with a (supposedly) soft heart; Chief Thunderthud, whose lines rotated around various exclamations of "Kowabunga!"; Princess Summerfall Winterspring; and Flubadub, a creature composed of a duck's head, a giraffe's neck, cocker-spaniel ears, a pig's tail, and a flowerpot hat. The father figure for this menagerie was Buffalo Bob, who acted as host-ringmaster, beginning each show by booming out at the Peanut Gallery, "Say kids! What time is it?" To which the forty assembled Peanuts would bellow back, "IT'S HOWDY DOODY TIME!!!!"

"Howdy Doody's" enormous appeal, particularly with the under-ten set, was reflected in hefty commercial terms. Not since Mickey Mouse in the 1930s, or radio heroes such as Buck Rogers, the Lone Ranger, or Charlie McCarthy, had there been such a marketing triumph; Howdy, in fact, made these earlier efforts look minimal. The merchandizing mills flooded stores with Howdy Doody paraphernalia, providing millions of grasping little hands with cowboy shirts, school bags, belts, dolls, earmuffs, piggy banks, umbrellas, and pinwheels.

"Howdy Doody" was the most popular children's show of its time and nurtured a generation of Baby Boomers from 1947 through the 1950s. Someone at NBC once estimated that, at the program's height of popularity, if all the kids on the waiting list for the Peanut Gallery stood in line waiting to get in, those at the end of the line would have finished college before their turn came.

There were no intellectual pretensions to "Howdy Doody"; mostly it was a lot of noise. But what Boomer can ever forget the delight of Clarabell's seltzer bottle?

Howdy Doody

National Museum of American History, Smithsonian Institution

The Lone Ranger

On January 30, 1933, The Lone Ranger made his first radio appearance on WXYZ in Detroit. Created by station owner George Trendle and written by Fran Striker, the Masked Rider of the Plains and his faithful companion, Tonto, hi-yoed on radio until 1954 and on television from 1949 until 1957.

The Lone Ranger was the sole survivor of an ambush against six Texas Rangers by Butch Cavendish and the Hole-in-the-Wall gang. The youngest Ranger, John Reid, was wounded; as he regained consciousness, he saw a shadowy figure move toward him:

TONTO: *Me . . . Tonto.*
REID: *What of the other Rangers? They were all my friends. One was my brother.*
TONTO: *Other Texas Rangers all dead. You only Ranger left. You lone Ranger now.*

And so the young Ranger became the knight rider, determined to right wrong for honor, God, and country. With Tonto he sought out each member of the Cavendish gang, stopping only for occasional rest and—more important—a fresh supply of silver bullets from a secret mine worked by a retired Ranger named Jim. The bullets were made of precious metal to remind the Lone Ranger to use them sparingly.

The Lone Ranger's character was based on a blend of the Zorro and Robin Hood legends, but in a wholesomely American rather than a swashbuckling way: a "guardian angel," as seen by his creators at WXYZ, who led the fight for law and order in the Old West. As the station announcer, Fred Foy, introduced the program, "Return with us now to those thrilling days of yesteryear! From out of the past, come the thundering hoofbeats of the great horse, Silver! The Lone Ranger rides again!"

Several actors preceded Brace Beemer—the best-known radio Ranger—as the Lone Ranger; he in turn was followed by Earl W. Graser. When Graser was killed in a car accident in 1941, Beemer took on the part again and played it until the last live radio broadcast (episode 2,956) on September 3, 1954.

TV's Lone Ranger (Clayton Moore) and Tonto (Jay Silverheels).

NBC

On television, Clayton Moore—with Jay Silverheels as Tonto—played the Lone Ranger for more than 180 episodes and is the actor most identified today with the role. A former trapeze artist who had played in movie serials, Moore was willing to perform half-hidden behind a mask; his face never was glimpsed by the camera.

From its inspired inception at WXYZ, the "Lone Ranger" has become part of America's folk pantheon.

Walt Disney 1901–1966

Walt Disney Productions was the first major movie studio to enter television with gusto. In the spring of 1954 Disney announced a twenty-six-week schedule for ABC, which would feature classic and new cartoons and films. It may have seemed a risky venture, but within three years "Disneyland" and the "Mickey Mouse Club" had proved to be brilliant financial successes; the TV shows in turn spent $10 million a year to advertise the new Disneyland amusement park and Disney movies.

"Disneyland" burst onto Wednesday-night television screens with such power that it blasted Arthur Godfrey away in the ratings race and captured the "biggest family audience in show business." In the first nine weeks that "Disneyland" was on the air in the fall of 1954, Disney stock soared. The center of this empire was known to acquaintances as a driving visionary with a touch of monomania. *TIME* magazine called him "a garage-type inventor with a wild guess in his eye and a hard pinch on his penny, a grass roots genius in the native tradition of Thomas A. Edison and Henry Ford."

Disney came to television as America's best-known animator, the creator of Mickey Mouse in the 1920s and of the first feature-length animated film, *Snow White and the Seven Dwarfs* (1937). He had won international acclaim as well for *Three Little Pigs* (1933), *Fantasia* (1940), *Pinocchio* (1940), *Bambi* (1942), and for such films using live characters as *Treasure Island* and *Robin Hood.* His pioneering television show, "Disneyland," introduced the phenomenally popular "Davy Crockett" series, filmed in color. Walt Disney Productions premiered the "Mickey Mouse Club" in 1955, giving American kids a new theme song ("M-I-C-K-E-Y M-O-U-S-E") and a penchant for wearing beanies surmounted by large rodent ears.

A friend once observed that Disney "built a whole industry out of daydreams." He had, above all, "the courage of his doodles."

The "Mickey Mouse Club," 1955.

Walt Disney with Tinkerbell.

Gene Autry born 1908

America's singing cowboy of screen, radio, and television, Gene Autry grew up on a small Texas ranch and sang in the local church choir. He was working the graveyard shift as a railroad telegraph operator one night when a stranger walked in to send a wire. Spotting the young man's mail-order guitar, the stranger said, "Boy, knock me off a tune." Gene did, and the man said, "You're wasting your time here. Why don't you get out and head for radio?" The stranger turned out to be Will Rogers, and young Autry took his advice and headed for New York.

This first venture failed, but then Autry and train dispatcher Jimmy Long wrote "That Silver-Haired Daddy of Mine," and his recording of this sentimental ballad launched his career. Sears hired him for a fifteen-minute program on WLS for thirty-five dollars a week; he also appeared on "National Barn Dance" and "National Farm and Home Hour."

His movie career began at Republic Studios, where *Tumbling Tumbleweeds* established an Autry genre of singing Westerns. The title derived from the picture's central Western ballad, Autry used his own name, and the cast numbered sidekicks such as Smiley Burnette and others from the "Barn Dance" show. The horse-opera troupe was completed with Autry's horse Champion, who had his own billing and received his own fan mail.

Autry made radio appearances in the late 1930s on the Eddie Cantor and Rudy Vallee shows and got his own program in the 1940s. When he enlisted in World War II, he took his oath over the air. After the war, he returned to radio on "Melody Ranch" and then moved to CBS television. Every Sunday at 7:00 P.M., the strains of "Back in the Saddle Again" wafted into America's living rooms, helping to spawn the children's Western craze on television. In 1951 Autry published his Cowboy Code of Ethics—a kind of Western Decalogue:

1. *A cowboy never takes unfair advantage, even of an enemy.*
2. *A cowboy never betrays a trust.*
3. *A cowboy always tells the truth.*
4. *A cowboy is kind to small children, to old folks, and to animals.*
5. *A cowboy is free from racial and religious prejudice.*
6. *A cowboy is always helpful, and when anyone's in trouble, he lends a hand.*
7. *A cowboy is a good worker.*
8. *A cowboy is clean about his person, and in thought, word and deed.*
9. *A cowboy respects womanhood, his parents and the laws of his country.*
10. *A cowboy is a patriot.*

Gene Autry

From *Great Radio Personalities in Historic Photographs* by Anthony Slide, Vestal Press, 1988

The "American Bandstand" regulars in action.

National Museum of American History, Smithsonian Institution

Dick Clark born 1929

Dick Clark began hosting "American Bandstand" in 1956. Broadcast from Philadelphia every afternoon for ninety minutes, "American Bandstand" had a daily audience estimated at forty million in 1959. From his perch atop the bandstand podium, Clark oversaw teenage regulars dancing to the music of early rockers like Buddy Holly, the Shirelles, and Chubby Checker—not to mention homegrown South Philly products like Frankie Avalon, Fabian, and Bobby Rydell. A plug of a new record by video jockey Clark on this show was said to guarantee 100,000 singles being sold before noon the next day.

Clark, after seeing a live broadcast of the Jimmy Durante-Garry Moore Show as a boy, decided "that's what I want to do." He worked as a disc jockey during college and got some early experience on television spinning records on "Paul Whiteman's TV Teen Club" before being chosen as host of "American Bandstand."

The format was simple. As Clark said, "I played records, the kids danced, and America watched." Top performers lip-synched their current hits, and Clark let teens on the set rate the records and then added his own judgment, as in, "I'd give this an 85. I like the beat, and it's easy to dance to." The teens who were regulars on the dance floor became minor celebrities themselves, receiving fan mail and popularizing such dances as the stroll, the Bristol stomp, and the twist. Some critics found it all too traumatic, but others saw Clark as "a symbol for all that's good in America's younger generation."

Among the performers who got their first national television exposure on "American Bandstand" were Bill Haley, James Brown, Jerry Lee Lewis, the Everly Brothers, and the Supremes.

Politics and the Media

The federal government's decision in the 1920s to allow radio to develop as a commercial enterprise rather than a government monopoly sounded the death knell for the smoke-filled back rooms of American politics. From early coverage of presidential speeches and national political conventions in radio's first decade, through Franklin Roosevelt's mastery of the media in the 1930s, to the downfall of Senator Joseph McCarthy in the 1950s, broadcasting cast a glaring spotlight on issues and personalities, and thereby exposed the political system. That the relationship between politics and the media was neither passive nor dispassionate was a natural outcome of the changed nature of American society: broadcasting created an ever-increasing demand for national political discussion, and those who best understood and used mass media's technology were those who succeeded. As radio and television provided the means for broadcasting the political message, the media itself became what historian Alan Brinkley has called "the central arbiter of political reality," conferring or rescinding legitimacy by the very power of the microphone's resonance or the camera's eye. The relationship between politics and the media was consummated with the Kennedy-Nixon debates in 1960, when the media-created debates themselves swung the balance in a campaign for the American presidency.

The implications of the media's active role in the nation's political life, particularly at the highest level, are increasingly debated today. Has broadcasting fulfilled its promise to act as a democratizing agent, or has it instead established homogenized standards that gain credibility because they prove palatable to the widest audience? To what extent has television elected the personality-celebrity—the media creation that Daniel Boorstin has termed "the human pseudo-event"—as the central figure in American politics?

Wendell Willkie, 1940 Republican nominee for President.

Watercolor, gouache, and india ink on paper by Al Hirschfeld. National Portrait Gallery, Smithsonian Institution

Franklin D. Roosevelt
1882–1945

Franklin D. Roosevelt's election as President in 1932 coincided with radio's own coming of age. At the touch of the dial, radio poured out a cascade of comedy, variety, and musical programming. Broadcasting—as a medium for entertainment, information, and increasing commercialization—had become a ubiquitous part of American life.

Not everyone found this presence reassuring. Critics railed against the growing commercial cacophony on the air, and in his 1934 *Our Master's Voice*, an angry James Rorty lambasted radio advertising: *It is like a grotesque, smirking gargoyle. . . . The gargoyle's mouth is the loudspeaker, powered by the vested interest of a two-billion dollar industry. . . . It is never silent. . . . For at least two generations Americans—the generations that grew up during the war and after—have listened to that voice as to an oracle. It has taught them how to live . . . how to be beautiful, how to be loved, how to be envied, how to be successful.*

And now the oracle became a central political presence. The 1932 presidential election was radio's quintessential campaign, contrasting not only two enormously different personalities, but two vastly disparate political points of view about how America would deal with the continuing Depression. Because radio had become a part of everyday life, people had instant access to issues and to the candidates themselves. FDR's personality was made for broadcasting; he communicated warmth, confidence, humor, and purpose, and did it in such a way that he might as well have been ensconced in listeners' living rooms, chatting before their own fireplaces. He was among the first to discover the impact of talking into the microphone as if he were being transported into the family circle. During the 1932 campaign, one New York critic described how FDR built "each word, each phrase, each sentence . . . with the invisible audience in mind." He visualized his listening audience not as some amorphous mass, but as small groups clustered around radios at home or in bars, restaurants, or cars. As a *New York Daily News* reporter noted, "While painting a verbal picture expansive enough for a museum mural, he reduced it to the proportions of a miniature hanging cozily on the wall of a living room."

Herbert Hoover lacked Roosevelt's graceful touch and was woefully remiss in such simple aspects as timing. A month before the election, Hoover made a radio broadcast that began at 8:30 P.M. *The Nation* described what started happening by 9:30 P.M.: *Listeners confidently awaited the President's concluding words. Confidently and also impatiently, for at 9:30 . . . Mr. Ed Wynn comes on the air. But Mr. Hoover had only arrived at point two of his twelve-point program. The populace shifted in its myriad seats; wives looked at husbands; children allowed to remain up until 10 o'clock on Tuesdays looked in alarm at the clock; 20,000 votes shifted to Franklin Roosevelt.*

As President, Roosevelt mastered the medium. His first "Fireside Chat" was on March 12, 1933, just over a week after his inauguration. In quiet, unpatronizing language, he explained the measures that he was taking to solidify the nation's banking system; Americans by the millions began making bank deposits again. Lillian Gish wrote to him that he seemed "to have been dipped in phosphorous," such was his incandescence on the radio.

Roosevelt timed his speeches exactly to fit their allotted radio slots. Once, while preparing a radio address at the White House, he stopped to say, "I'll pause here. That'll get 'em." He even joked, "I know what I'll do when I retire. I'll be one of these high-powered commentators."

At his death *Broadcasting* magazine listed the ways in which radio had become "the anchor political medium" during Roosevelt's tenure: "News releases were timed for radio deadlines. . . . Press conferences became 'radio press conferences.' . . . Radio galleries sprouted in Congress. Radio correspondents were accredited on the press level at home and in the war theatres." Radio's Golden Age paralleled Roosevelt's presidency—and he, by his facility on the air, helped the mass medium revolutionize politics.

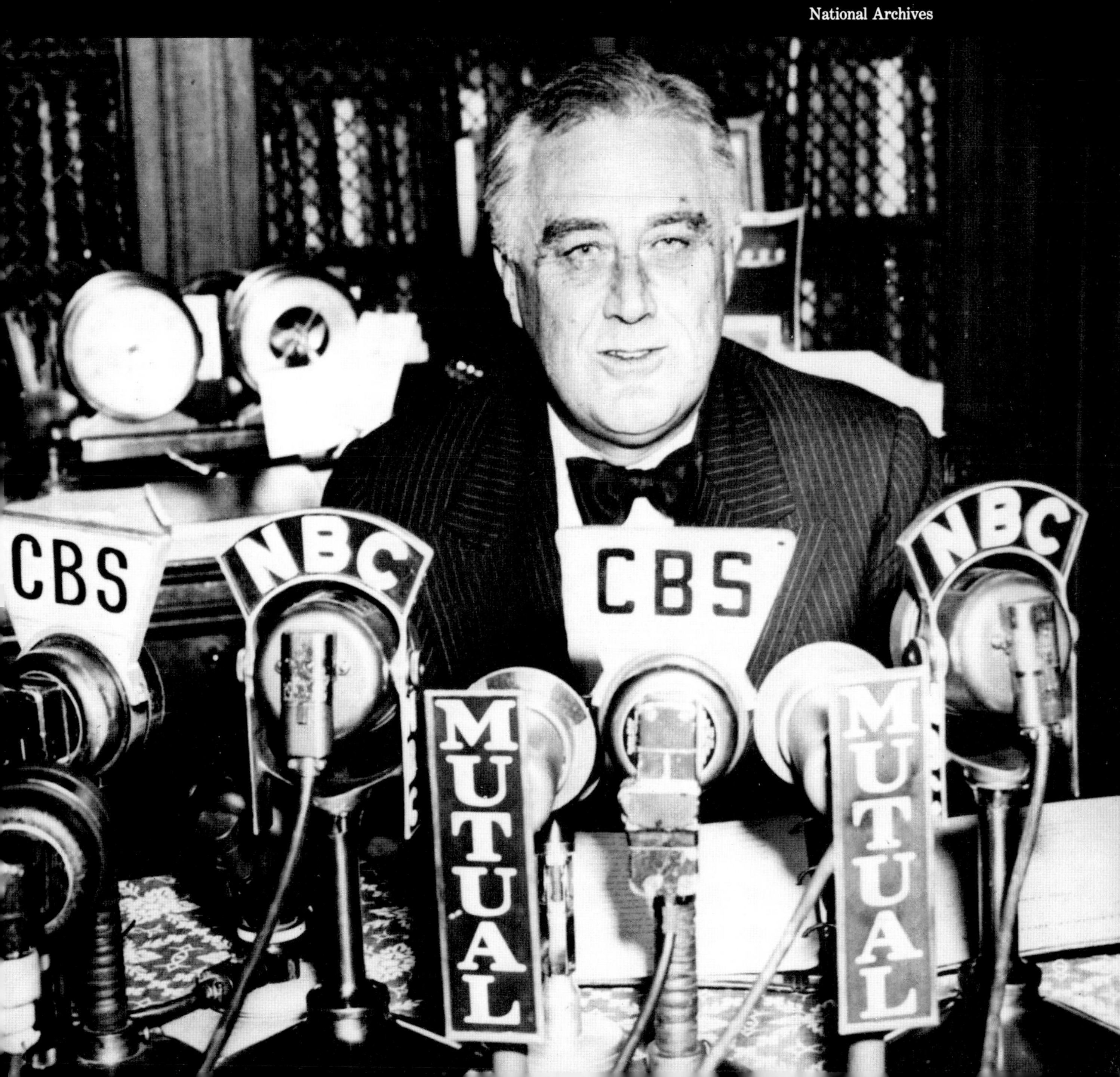

Franklin D. Roosevelt, whose mastery of radio helped revolutionize American politics.

National Archives

Charles E. Coughlin 1891–1979

In 1928 Senator Pat Harrison of Mississippi declaimed, "Radio will purify politics. The blatant demagogue once sought the little hamlet where not even a newspaper account of his speech would be circulated. . . . Radio has come to do away with that. The venomous darts or scurrilous anchors cannot pass through the air."

Harrison was too optimistic. The problem with demagogues is that they often appear in some credible guise, establish a following from a ground swell of popularity, and feed that popularity by tapping into some latent hope or fear.

In the case of Father Charles E. Coughlin, the guise was a collar of credibility. Assigned in 1926 to a small church in suburban Detroit, the Church of the Little Flower, Coughlin began broadcasting over WJR in the hope of enlarging his parish. His first broadcasts were aimed at children, but as he began fixing on political issues—against the Communist "red serpent," for a Populist remonetization of silver, and against the wealthy capitalists—he was barraged with sympathetic letters and contributions. He organized a Radio League of the Little Flower with a one-dollar-per-year membership, and as the Depression deepened, so did the extent of his support. He came to national attention in the winter of 1931–1932, when he attacked Herbert Hoover and the international bankers on his "Golden Hour" broadcast over CBS. The network told him to stop such political attacks and to submit scripts in advance; he did for his next program but then used his air time to tell listeners about the CBS "muzzle." As a result the network received 1.25 million letters of protest.

Coughlin was eased off CBS and shut out by NBC as well, but he was able to find time on a group of stations that included WOR in New York. Throughout 1932 he campaigned for Franklin Roosevelt, characterizing himself as the "radio priest" who spoke up for the common man. By 1934 *Fortune* magazine named him "just about the biggest thing that ever happened to radio," and a WOR poll found him the "most useful citizen" in America.

Later that year Coughlin organized the National Union for Social Justice, which many viewed as a proto-Fascist party. Then, fighting for remonetization of silver, Coughlin got into a fight with Roosevelt's secretary of the treasury, Henry Morgenthau, over administration attempts to block such legislation in Congress. Coughlin blasted Morgenthau and Wall Street and praised silver as a "gentile" metal.

Converging with Coughlin's proto-Fascist lobby was the emergence of Huey Long and his Share-the-Wealth movement, which by 1935 had established clubs in eight thousand communities and had an estimated membership of seven million. If, as rumor had it, Coughlin and Long were to merge their movements, the Democrats faced possible catastrophe in 1936. Long's assassination in Baton Rouge stanched that merger; Coughlin called the death of the "Kingfish" (Long's nickname) "the most regrettable thing in modern history."

Coughlin had spotlighted radio's double-edged political potential—not only to purge politics but to bolster demagoguery. A former New Dealer said: "You can laugh at Huey Long—you can snort at Fr. Coughlin—but this country was never under a greater menace . . . with the radio and the newsreel to make them effective."

Coughlin reached his peak in 1939, when he called for his listeners to organize "an army of peace" to march on Washington, D.C., in protest of the liberalization of the neutrality laws. As Coughlin's reputation as a disciple of fascism and a purveyor of hate and violence grew, the National Association of Broadcasters began to clamp down on him. By the summer of 1940 enough stations had refused to renew Coughlin's contract that he was finally off the air. But it had been a frightening flirtation.

Father Coughlin

Lithograph by Hugo Gellert, circa 1936. National Portrait Gallery, Smithsonian Institution

Photograph of Joseph McCarthy taken from the television screen during the Army-McCarthy hearings, 1954.

CBS Inc.

Joseph McCarthy 1909–1957

Television reached a political coming of age with the Army-McCarthy hearings in 1954. For thirty-six days the hearings documented not so much alleged Communist subversion, but the self-destruction of Senator Joseph McCarthy. The television camera ruthlessly recorded the tactics of the chairman of the Senate Government Operations Committee's subcommittee on investigations and contributed to his censure by the Senate on December 1, 1954.

The junior Senator from Wisconsin was first elected to the United States Senate in 1946. Reelected in 1952, he had gained worldwide attention by a speech in Wheeling, West Virginia, in February 1950, in which he claimed to have a list of 205 people "that were known to the secretary of state as being members of the Communist Party and who nevertheless are still working and shaping the policy of the State Department." The speech provoked headlines across the country, and though the infamous list was never produced, McCarthy was launched as a leading red-baiter.

Basing his anti-Communist campaign on innuendo and fear rather than hard evidence, McCarthy generated such an atmosphere of zealous "patriotism" that few braved confronting him. Edward R. Murrow in 1950 said that the "whole McCarthy business is squalid beyond words."

Murrow and coproducer Fred W. Friendly told their "See It Now" staff in 1953 to begin collecting all available McCarthy footage. On March 9, 1954, they broadcast "a report on Senator Joseph R. McCarthy," composed largely of news footage and Murrow's brief commentary. Mostly, McCarthy was just shown talking. Friendly was reportedly so nervous at the beginning of the program that he couldn't even start his stopwatch to time the show. At the program's conclusion, Murrow observed: *The actions of the junior Senator from Wisconsin have caused alarm . . . and whose fault is that? Not really his. He didn't create this situation of fear; he merely exploited it, and rather successfully. Cassius was right: "The fault, dear Brutus, is not in our stars but in ourselves. . . . " Good night, and good luck.*

In April 1954 the Senate subcommittee on investigations began the hearings that would show, day after day, the conflicting testimonies of McCarthy, the Secretary of the Army, and their aides. As the hearings wore on, television's power to convey an atmosphere of vivid immediacy—and to lay personality bare—became clear. Almost three-quarters of America's population watched the hearings at some point. As a CBS psychologist later analyzed, viewers came to see the principals as people whom they actually knew, and made their judgments on the level of face-to-face contact. One housewife was reported to have changed her mind about supporting McCarthy after seeing him in daily action: "I just started to know more about him really . . . and I became afraid of such a man, that the power he had was terrible to make other men feel uncomfortable."

The Army-McCarthy hearings were more than a political spectacle. They showed that television could create or destroy with equal authority.

Kennedy-Nixon Debates
John F. Kennedy 1917–1963
Richard M. Nixon born 1913

The 1960 presidential campaign between John F. Kennedy and Richard M. Nixon was a turning point in the relationship between television and American politics: television moved from the wings to center stage, both in its vigilant documenting of the daily campaigns and in the spectacle it itself created with the series of "Great Debates" between Kennedy and Nixon.

Family watching the Kennedy-Nixon debate on TV.

National Archives

CBS president Frank Stanton paved the way for the debates, convincing Congress to set aside by joint resolution the "equal time" rule of the Federal Communications Act. On September 26, 1960, the first debate was held at Chicago's WBBM. Moderated by Howard K. Smith, the debate began with opening statements, followed by questions from newsmen (Robert Fleming of ABC, Stuart Novins of CBS, Sander Vanocur of NBC, and Charles Warren of Mutual Radio), comments by the candidates on each other's answers, and summations. An estimated 115 million people listened to at least one of the four debates. There is no question that, in this marginally close election, television was the determining factor. For example, in surveys conducted by the Sindlinger research company, it was shown that before the first debate, 39.3 percent of those polled thought that Nixon would win, to Kennedy's 31.2 percent. The day after the fourth debate, Kennedy outpolled Nixon 33 to 29 percent. But of those who listened only on radio, 48.7 percent picked Nixon and only 21 percent chose Kennedy.

Kennedy always credited television with giving him the edge in the election, once saying, "We'd never have won without that gadget." Despite its all-seeing eye, TV is not passive in its depiction of personality: the camera focuses on the character traits of the candidates—credibility, confidence, warmth, even their entertainment value—as much as it does on political points of view. More than ever before, under the magnifying glass of television, how something was said had become as important as what was said. Certainly since the Kennedy-Nixon debates, the face of American politics has never been the same.

The first Kennedy-Nixon debate, broadcast from WBBM in Chicago on September 26, 1960.

CBS Inc.

Selected Bibliography

Allen, Fred. Papers. Boston Public Library, Massachusetts.

———. *Treadmill to Oblivion*. Boston: Little, Brown, 1954.

Allen, Frederick Lewis. *Since Yesterday: The 1930s in America*. New York: Harper, 1939.

———. *Only Yesterday: An Informal History of the Nineteen-Twenties*. New York and London: Harper, 1951.

Allen, Robert C. *Speaking of Soap Operas*. Chapel Hill and London: University of North Carolina Press, 1985.

Archer, Gleason L. *History of Radio to 1926*. New York: The American Historical Society, 1938.

Bannerman, R. LeRoy. *Norman Corwin and Radio: The Golden Years*. University, Ala.: University of Alabama Press, 1986.

Banning, William Peck. *Commercial Broadcasting Pioneer: The WEAF Experiment, 1922–1926*. Cambridge, Mass.: Harvard University Press, 1946.

Barnouw, Erik. *A History of Broadcasting in the United States*. 3 vols. New York: Oxford University Press, 1966–1970.

Benny, Jack. Papers. American Heritage Center, The University of Wyoming, Laramie.

———. Papers. University of California, Los Angeles.

Berg, Gertrude, with Cherney Berg. *Molly and Me*. New York: McGraw-Hill, 1961.

Berle, Milton. *B.S. I Love You: Sixty Funny Years with the Famous and the Infamous*. New York: McGraw-Hill, 1988.

Bilby, Kenneth. *The General: David Sarnoff and the Rise of the Communications Industry*. New York: Harper & Row, 1986.

Bliss, Edward, Jr., ed. *In Search of Edward R. Murrow, 1938–1961*. New York: Alfred A. Knopf, 1967.

Bliven, Bruce. "How Radio Is Remaking Our World." *Century* 108 (June 1924).

Blum, Daniel C. *Pictorial History of TV*. Philadelphia: Chilton, 1958.

Blythe, Cheryl, and Susan Sackett. *Say Goodnight, Gracie!* New York: E. P. Dutton, 1986.

Boorstin, Daniel J. *The Image: A Guide to Pseudo-Events in America*. New York: Atheneum, 1971.

Buxton, Frank, and Bill Owen. *Radio's Golden Age: The Programs and the Personalities*. New York: Easton Valley Press, 1966.

Campbell, Robert. *The Golden Years of Broadcasting: A Celebration of the First 50 Years of Radio and TV on NBC*. New York: Charles Scribner's Sons, 1976.

Cantril, Hadley, with the assistance of Hazel Gaudet and Herta Herzog. *The Invasion from Mars: A Study in the Psychology of Panic*. Princeton, N.J.: Princeton University Press, 1940.

Chase, Gilbert. *Music in Radio Broadcasting*. New York: McGraw-Hill, 1946.

Chayefsky, Paddy. Papers, 1937–1972. State Historical Society of Wisconsin, Madison.

Clark Collection. Archives Center, National Museum of American History, Smithsonian Institution, Washington, D.C.

Correll, Charles J., and Freeman F. Gosden. *All About Amos 'n' Andy*. New York: Rand McNally, 1929.

Corwin, Norman. *Thirteen by Corwin*. New York: Holt, 1942.

———. *More by Corwin: 16 Radio Dramas*. New York: Holt, 1944.

Crosby, Bing, as told to Peter Martin. *Call Me Lucky*. New York: Simon & Schuster, 1953.

Crosby, John. *Out of the Blue*. New York: Simon & Schuster, 1952.

Czitrom, Daniel J. *Media and the American Mind*. Chapel Hill: University of North Carolina Press, 1982.

Daly, John Charles. Papers, 1935–1967. State Historical Society of Wisconsin, Madison.

De Forest, Lee. Diary. Manuscript Division, Library of Congress, Washington, D.C.

They don't call them variety shows for nothing. One of the best was "The Garry Moore Show."

CBS Inc.

———. *Father of Radio: The Autobiography of Lee De Forest.* Chicago: Wilcox and Follett, 1950.

DeLong, Thomas A. *The Mighty Music Box: The Golden Age of Musical Radio.* New York: Hastings House, 1980.

———. *Pops: Paul Whiteman, King of Jazz.* Piscataway, N.J.: New Century Publishers, 1983.

Douglas, Susan J. *Inventing American Broadcasting, 1899–1922.* Baltimore, Md., and London: The Johns Hopkins University Press, 1987.

Dumont Collection. Archives Center, National Museum of American History, Smithsonian Institution, Washington, D.C.

Durstine, Roy. "We're on the Air." *Scribner's* 83 (May 1928).

Fein, Irving A. *Jack Benny: An Intimate Biography.* New York: Pocket Books, 1976.

Fox, Richard Wightman, and T. J. Jackson Lears. *The Culture of Consumption: Critical Essays in American History, 1880–1980.* New York: Pantheon Books, 1982.

Friendly, Fred W. *Due to Circumstances Beyond Our Control* . . . . New York: Random House, 1967.

Galanoy, Terry. *Tonight!* Garden City, N.Y.: Doubleday and Company, 1972.

Gitlin, Todd. *Inside Prime Time.* New York: Pantheon Books, 1983.

Gross, Ben. *I Looked and Listened: Informal Recollections of Radio and TV.* New York: Random House, 1954.

Harmon, Jim. *The Great Radio Heroes.* Garden City, N.Y.: Doubleday, 1967.

Hart, Joseph R. "Radiating Culture." *Survey* 47 (March 18, 1922).

Haskins, James, with Kathleen Benson. *Nat King Cole.* New York: Stein and Day, 1984.

Hoopes, James. *Van Wyck Brooks: In Search of American Culture.* Amherst: University of Massachusetts Press, 1977.

Horowitz, Joseph. *Understanding Toscanini.* New York: Alfred A. Knopf, 1987.

Innis, Harold A. *Empire and Communications.* Oxford, U.K.: Clarendon Press, 1950.

Kaempffert, Waldemar. "The Social Destiny of Radio." *Forum* 71 (June 1924).

Kaltenborn, Hans V. Papers, 1883–1964. State Historical Society of Wisconsin, Madison.

———. *I Broadcast the Crisis.* New York: Random House, 1938.

Kasson, John F. *Civilizing the Machine.* New York: Grossman, 1976.

Kirby, Edward M., and Jack W. Harris. *Star-Spangled Radio.* Chicago: Ziff-Davis Publishing Company, 1948.

Koppes, Clayton R. "The Social Destiny of Radio: Hope and Disillusionment in the 1920s." *South Atlantic Quarterly* 68 (1964).

Kovacs, Ernie. Papers. University of California, Los Angeles.

Lazarsfeld, Paul F., and Frank N. Stanton, eds. *Radio Research 1941.* New York: Duell, Sloan & Pearce, 1941.

Lyons, Eugene. *David Sarnoff—A Biography.* New York: Harper, 1966.

MacDonald, J. Fred. *Don't Touch that Dial! Radio Programming in American Life from 1920 to 1960.* Chicago: Nelson-Hall, 1979.

MacLeish, Archibald. *The Fall of the City: A Verse Play for Radio.* New York: Farrar & Rinehart, 1937.

McLuhan, Marshall. *Understanding Media: The Extensions of Man.* New York: McGraw-Hill, 1964.

McMahon, Morgan E. *A Flick of the Switch, 1930–1950.* Palos Verdes Peninsula, Calif.: Vintage Radio, 1976.

McNamee, Graham, in collaboration with Robert Gordon Anderson. *You're on the Air.* New York and London: Harper, 1926.

Marc, David. *Demographic Vistas: Television in American Culture.* Philadelphia: University of Pennsylvania Press, 1984.

Marchand, Roland. *Advertising the American Dream: Making Way for Modernity, 1920–1940*. Berkeley: University of California Press, 1985.

Marquis, Alice Goldfarb. *Hopes and Ashes: The Birth of Modern Times, 1929–1939*. New York: The Free Press, 1986.

Marx, Groucho. Miscellaneous Collection. Archives Center, National Museum of American History, Smithsonian Institution, Washington, D.C.

Marx, Leo. *The Machine in the Garden: Technology and the Pastoral Ideal in America*. New York: Oxford University Press, 1964.

Murrow, Edward R. *This Is London*. New York: Simon & Schuster, 1941.

National Broadcasting Company. Records, 1921–1969. State Historical Society of Wisconsin, Madison.

O'Connor, John E., ed. *American History/American Television: Interpreting the Video Past*. New York: Frederick Ungar Publishing Co., 1983.

Osgood, Dick. *WYXIE Wonderland: An Unauthorized 50-Year Diary of WXYZ Detroit*. Bowling Green, Ohio: Bowling Green University, Popular Press, 1981.

Paper, Lewis J. *Empire: The Life and Times of William S. Paley*. New York: St. Martin's, 1987.

Persons, Stow. *The Decline of American Gentility*. New York and London: Columbia University Press, 1973.

Peyser, Joan. *Bernstein, a Biography*. New York: William Morrow, 1987.

Radio Annual. New York: Radio Daily, annual, 1938–1960.

Rhymer, Paul. Papers, 1928–1972. State Historical Society of Wisconsin, Madison.

Rorty, James. *Our Master's Voice: Advertising*. New York: John Day, 1934.

Rosenberg, Bernard, and David Manning White. *Mass Culture: The Popular Arts in America*. Glencoe, Ill.: The Free Press, 1956.

Saudek, Robert. "Program Coming in Fine, Please Play Japanese Sandman." *American Heritage* 16, no. 5 (August 1965).

Seldes, Gilbert. *The Great Audience*. New York: Viking, 1950.

———. *The Public Arts*. New York: Simon & Schuster, 1956.

Serling, Rod. Papers, 1943–1971. State Historical Society of Wisconsin, Madison.

Settel, Irving. *A Pictorial History of Radio*. New York: Bonanza Books, 1960.

Shales, Tom. *On the Air!* New York: Summit Books, 1982.

Shirer, William L. *Berlin Diary: The Journal of a Foreign Correspondent, 1934–1941*. New York: Alfred A. Knopf, 1941.

Smith, Kate. Miscellaneous Collection. Boston University, Massachusetts.

Smulyan, Susan Renee. " 'And Now a Word From Our Sponsors . . .': Commercialization of American Broadcast Radio, 1920–1934." Ph.D. diss., Yale University, 1985.

Spalding, John W. "1928: Radio Becomes a Mass Medium." *Journal of Broadcasting* 8 (1963–1964).

Sperber, A. M. *Murrow: His Life and Times*. New York: Freundlich Books, 1986.

Spiller, Robert E., and Eric Larrabee. *American Perspectives: The National Self-Image in the Twentieth Century*. Cambridge, Mass.: Harvard University Press, 1961.

Striner, Richard A. "The Machine as Symbol: 1920–1939." Ph.D. diss., University of Maryland, 1982.

Sullivan, Ed. Papers, 1920–1974. State Historical Society of Wisconsin, Madison.

Summers, Harrison B., ed. *A Thirty-Year History of Programs Carried on National Radio Networks in the United States, 1926–1956*. Columbus: Ohio State University, 1958.

Susman, Warren I. *Culture as History: The Transformation of American Society in the Twentieth Century*. New York: Pantheon Books, 1984.

Tebbel, John. *The Media in America*. New York: Crowell, 1974.

Thurber, James. "Soapland." In *The Beast in Me and Other Animals*. New York: Harcourt, Brace, 1948.

Trachtenberg, Alan. *The Incorporation of America*. New York: Hill & Wang, 1982.

Ulanov, Barry. *The Incredible Crosby*. New York and Toronto: Whittlesey House, 1948.

Weibe, Robert H. *The Search for Order*. New York: Hill & Wang, 1967.

White, Paul W. *News on the Air*. New York: Harcourt, Brace, 1947.

Wilson, Richard Guy, Dianne H. Pilgrim, and Dickran Tashjian. *The Machine Age in America, 1918–1941*. New York: Harry N. Abrams, 1986.

Index

Italicized page numbers refer to illustrations.

Photography Credits

Sheldan Collins: pp. 7, 15, 22, 68, 79, 89 (Jack Paar by Al Hirschfeld), 131, 133, 173
Wayne David Geist: p. 76
Eugene Mantie/Rolland White: cover illustration and pp. 12, 30, 33 (David Sarnoff by S. J. Woolf), 39, 63 (Will Rogers by Jo Davidson), 82, 98, 103, 105, 146 (Archibald MacLeish by Milton Hebald), 189, 199 (*The Corpse in the Living Room* by Boris Chaliapin)
Laurie Minor: p. 28

The October 26, 1959, cover of TIME *magazine* (The Corpse in the Living Room) *featured TV's top guns (clockwise from left): Peter Gunn (Craig Stevens), Richard Diamond (David Janssen), Philip Marlowe (Philip Carey), crook, Perry Mason (Raymond Burr), and Stu Bailey (Efrem Zimbalist, Jr.).*

Gouache on board by Boris Chaliapin, 1959. National Portrait Gallery, Smithsonian Institution; gift of Time, Inc.

Dick Clark manning the "American Bandstand" podium.

National Museum of American History, Smithsonian Institution

On the Air was designed by Polly Sexton of Washington, D.C., and typeset in Century Expanded by BG Composition, Inc. of Baltimore, Maryland. It was printed on Warren's Lustro Offset Enamel dull paper by Collins Lithographing and Printing Company, Inc. of Baltimore, Maryland.

ROUD WINS
WALTER O'KEEFE
MILTON CROSS
DON WILSON
JIMMY FIDLER
ALEXANDER WOOLLCOTT
ED WYNN
GRAHAM McNAMEE
UNCLE EZRA
SINGING LADY
PICK &
PAT
JESSICA DRAGONETTE
ROBERT L. RIPLEY
LOWELL THOMAS
MOLLY
JACK OAKIE
JOHN CHARLES THOMAS
GENERAL HUGH JOHNSON
CHARLES BUTTERWORTH
ALFRED WALLENSTEIN
LAWRENCE TIBBETT
EDWIN C. HILL
ABNER
PROF QUIZ
TED HUSING
DICK POWELL
PAUL WHITEMAN
KENNY BAKER
DEANNA DURBIN
MARTHA RAYE
STOOPNAGLE &
BUDD
CAB CALLOWAY
NELSON EDDY
FRANCES LANGFORD
JEANETTE MacDONALD
ROBERT TAYLOR
LANNY ROSS
PHIL BAKER
GUY LOMBARDO
DUCK
DON AMECHE
AL PEARCE
FANNY BRICE
BOB BURNS
TYRONE POWER
ONALD
MARCH of TIME
BEATRICE LILLIE
JOE PENNER
MAGIC KEY

Key for Radio Talent: *detail illustrated on back cover*
Illustrated in full on page 12